# Modern
# Conformational
# Analysis

# Methods in Stereochemical Analysis

Series Editor
Alan P. Marchand, Denton, Texas, USA

Advisory Board
A. Greenberg, Charlotte, North Carolina, USA
I. Hargittai, Budapest, Hungary
J. Liebman, Baltimore, Maryland, USA
E. Lippmaa, Tallinn, Estonia
Leo A. Paquette, Columbus, Ohio, USA
Paul von Ragué Schleyer, Erlangen, Germany
S. Sternhell, Sydney, Australia
Y. Takeuchi, Tokyo, Japan
F. Wehrli, Philadelphia, Pennsylvania, USA
D. H. Williams, Cambridge, UK
N. S. Zefirov, Moscow, Russia

# Modern Conformational Analysis

## Elucidating Novel Exciting Molecular Structures

*Helena Dodziuk*

Helena Dodziuk
Institute of Organic Chemistry
Polish Academy of Sciences
01-224 Warsaw, Kasprzaka 44
Poland

This book is printed on acid-free paper. ∞

**Library of Congress Cataloging-in-Publication Data**

Dodziuk, Helena.
 Modern conformational analysis : elucidating novel exciting
molecular structures / by Helena Dodziuk.
   p.    cm. — (Methods in stereochemical analysis)
 Includes bibliographical references and index.
 ISBN 1-56081-689-9 (alk. paper)
 1. Stereochemistry.  2. Conformational analysis.  I. Title.
II. Series.
QD481.D812   1995
547.12—dc20                                    95-16397
                                                    CIP

© 1995 VCH Publishers, Inc.

Printed in the United States of America

ISBN 1-56081-689-9 VCH Publishers, Inc.


Printing History:
10 9 8 7 6 5 4 3 2 1

Published jointly by

VCH Publishers, Inc.          VCH Verlagsgesellschaft mbH     VCH Publishers (UK) Ltd.
220 East 23rd Street          P.O. Box 10 11 61               8 Wellington Court
New York, New York 10010      69451 Weinheim, Germany         Cambridge CB1 1HZ
                                                              United Kingdom

# Preface

Because of the rapid development of organic syntheses and that of theoretical and experimental methods of structure elucidation, numerous organic molecules with unusual spatial structure have been synthesized in recent years. Their captivating shapes and the corresponding charge distributions determine molecular properties and the interactions between molecules. Thus they allow a better understanding of the phenomenon of self-organization and form a foundation for comprehension of the origin of life. Complementarity of the shapes of interacting molecules is also indispensable for a rationalization of pharmacologic activity of drugs and for replacing complicated multistep syntheses with much simpler and efficient one-pot reactions.

After a short introduction, the presentation will start with a discussion of physical methods as a source of information on molecular spatial structure. Owing to the enormous diversity of science today, most monographs in physical organic chemistry deal with a specific physical method in depth. As a result, a comparison of structural determination methods and the complementarity of the information they provide have not received sufficient attention. Therefore, for the most important structure determination methods basic principles of their operation will be outlined. This will be followed by a brief description of molecular structure information and the energetics the method provides. Then the accuracy and complementarity of various methods will be discussed in more detail.

Theoretical calculations and the interaction between theory and synthetic chemistry form another interesting aspect of the recent rapid development of organic chemistry that is presented. Sometimes, as happened with the now-famous $C_{60}$, theoretical ideas appear too early to gain recognition. In other cases theoretical and synthetic works stimulate each other. An example of a fruitful cooperation between theoretical and synthetic chemistry provides the prediction of existence and properties of [1.1.1]propellane, which, after a successful synthesis of this molecule, enabled theoreticians to give a new definition of a chemical bond. Recent calculations on molecules that should possess a tetravalent carbon atom with a pyramidal arrangement of substituents will undoubtedly lead to the synthesis of such unusual systems in the near future. A brief discussion of molecular symmetry illustrated by numerous examples of beautiful, symmetrical molecules recently synthesized will follow the chapter on theoretical calculations. Then the importance of molecular chirality will be stressed, and the Cahn, Ingold, and Prelog classification of chirality and its modification will be presented.

A short review of geometrical parameters of typical organic molecules will be followed by a discussion of new molecules having unusual three-dimensional structure. The latter group will include:

1. Hydrocarbons that have tetravalent carbon atoms with a spatial arrangement of substituents different considerably from the tetrahedral one;
2. Unsaturated and aromatic hydrocarbons with a nonplanar arrangement of bonds around trivalent carbon atoms;
3. Molecules of novel topology mimicking Möbius strips, knots, etc., and some other unusual molecules; and
4. Supramolecular systems exemplified mainly by cyclodextrins and their complexes.

A short discussion of the aggregation and self-aggregation phenomena that are crucial in biological systems and will find various applications in chemistry and industry will be also given in this chapter.

Chemists began to look at molecules as three-dimensional objects not long ago, and the importance of their spatial organization for molecular properties only now begins to gain due recognition. Molecular modeling has developed recently on the basis of simple molecular mechanics, molecular dynamics, and quantum calculations, using the sophisticated graphic capabilities of PCs and more powerful computers. The technique called *computer-assisted* (or *-aided*) *molecular design* (CAMD) is used to carry out model calculations and to produce beautiful images of molecular models, allowing one to accelerate the industrial development of new materials with predetermined desired properties. An example of CAMD applications in drug design will conclude the chapter on molecular modeling. The application of organic molecules as drugs and other, mostly prospective, applications of organic materials such as conductors, semiconductors, magnets, and nonlinear optical materials will be briefly reviewed in the last chapter.

This book is written for organic and physical chemists who are acquainted with classical conformational analysis and want to learn more about recent developments and perspectives in this field. They know that, as a rule, cyclohexane assumes a chair conformation, but they may find it interesting that its six-membered ring can in certain cases be planar. For most other molecules discussed in this book the aim will be just the opposite, that is, to show that planar molecular formulas are crude representations of complex three-dimensional objects and that their shape and charge distribution determine molecular properties. This book encompasses a very wide area presented in an introductory way. The chapters are written as separate entities; therefore, some of them can be skipped on a first reading. The comprehensive reference lists should enable an interested reader a deeper insight into the problems he or she wants to learn in detail.

As a physicist working in the domain of organic stereochemistry, I am fascinated by the beauty and variety of molecular shapes. To present some of them and to convey some of my fascination to the readers are the main aims of this book.

The creation of this book would not be possible without the help of several people. My thanks are due to Professor J. Jurczak and Professor L. Stefaniak for their constant support and encouragement. The whole manuscript was read and critically reviewed by Professor Z. R. Grabowski, Professor A. Zamojski, and Dr. W. Kozminski. I am indebted to them for their critical remarks. I am also grateful to Professor J. Lipkowski for reading and critically commenting on Chapters 2 and 10. Chapter 3 would be completely different without the critical comments of Professor M. Jaszuński. Critical remarks on Chapter 11 and photographs by Dr. Nowiński also have to be gratefully acknowledged. The NMR spectra presented in Chapter 2 were recorded by Mr. J. Sitkowski. M.Sc., to whom I am also very obliged. My thanks are also due to those mentioned in figure captions who provided me with drawings and photographs reproduced in this book. I believe that all the comments and remarks have considerably improved this book, but I am the only one to be blamed for any of its faults. There is a plethora of computer programs available on the market. Those mentioned in this book are those known to the author, and their mentioning does not mean that they are superior to the others.

Helena Dodziuk
Warsaw, Poland
April 1995

# Contents

# 1

# Introduction

## 1.1 Initial Remarks

In his pioneering works [1] in the middle of nineteenth century, Louis Pasteur recognized that different behaviors of the isomers of tartaric acid **1–3** (see Section 1.2 for the convention used in these formulae) were due to the differences in their three-dimensional structures. Twenty years later the van't Hoff [2] and LeBel [3] hypothesis on the tetrahedral arrangement of substituents around a tetravalent carbon atom laid the foundation of organic stereochemistry. The idea met fierce criticism, sometimes expressed in a style far from that usually encountered in scientific discussions. A respectable German professor, Adolf Wilhelm Hermann Kolbe, must have forgotten his *Kinderstube* [4] when writing "A Dr. J. van't Hoff, of the Veterinary School of Utrecht, has no liking, apparently, for exact chemical investigation. He has considered it more comfortable to mount Pegasus (apparently borrowed from the Veterinary School) and to proclaim in his 'La Chimie dans l'éspace' how the atoms appear to him to be arranged in space, when he is on the chemical Mt. Parnassus which he has reached by bold fly." Contrary to Kolbe's intentions, such a stormy attack had awakened considerable interest in the van't Hoff ideas, thus helping in their acceptance. However, it took more than 50 years fully to recognize the importance of the spatial arrangement of atoms in molecules for their properties. Molecular geometry is described by the directionality of bonds linking its constituent atoms. This directionality, in turn, can be described in terms of the hybridization of electron orbitals for valence electrons of carbon atom. There are three types of hybridization for carbon atoms corresponding to three spatial configurations of these orbitals around an atomic nucleus. In the case of the $sp^3$ configuration (see the discussion of hybridization in Section 3.2), the regions with the highest electron density are grouped around four axes pointing toward every second vertex of a cube with the carbon atom in its center. These axes show the directions of four $\sigma$ bonds formed by the central carbon atom and, consequently, the positions of four substituents on the atom. $sp^2$ hybridization corresponds to a system with a carbon atom forming three $\sigma$ bonds lying in one plane and one $\pi$ bond perpendicular to the plane. The last $sp$ configuration describes a linear configuration of substituents on the corresponding carbon atom, with one $\sigma$ and two mutually perpendicular $\pi$ bonds. In such a way the spatial structure of organic molecules depends on the hybridization of the electronic orbitals of its constituent atoms. Thus the electronic structure of atoms forming a molecule determines its geometry.

|  | COOH |  | COOH |  | COOH |  |  | COOH |  |
|---|---|---|---|---|---|---|---|---|---|
| HO — | — H | H — | — OH | H — | — OH |  | HO — | — H |  |
| H — | — OH | HO — | — H | H — | — OH |  | HO — | — H |  |
|  | COOH |  | COOH |  | COOH |  |  | COOH |  |
|  | **1** |  | **2** |  | **3** |  |  |  |  |

Today structure–activity relationships using molecular modeling are increasingly used to predict the biological activity of drugs in pharmaceutical companies or to create organic conductors, semiconductors, and molecular switches. On the other hand, a different approach to planning organic reactions is on the way. Until recently, the only way to synthesize a complicated organic molecule was to plan a series of reactions such as that leading to a significant local hormone, prostaglandin $PGF_2\alpha$, presented in Scheme 1.1 [6]. Such multistep reactions are very time consuming and have low overall yields, since the yield is equal to the product of the yields for all reactions constituting the series. Even if the yield were as high as 90% for every reaction in the scheme, the effective yield for the whole series would be less than 15%.

In contrast to Scheme 1.1, the reactions shown in Schemes 1.2 [7] and 1.3 [8] proceed in one step, producing complicated molecules from simple reagents thanks to molecular self-organization, governed by their shapes and electron density distributions. Today it is not possible to synthesize a complicated natural product in such so-called "one-pot" reactions. However, several steps in the prostaglandin synthesis have been combined, speeding up the reactions and enhancing the total yield [9]. Progress will certainly take place in this direction, and one-pot reactions will inevitably be in wide use in the future. Other examples of reactions of self-organizing components will be discussed in Sections 9.1 and 10.5. At present we only begin to understand how such reactions proceed, but the ultimate aim of their studies is prediction, enabling a proposal of such one-pot multistep reactions.

The concept of self-assembly started with biochemical studies on tobacco mosaic virus [10–12] and the enzyme ribonuclease [13, 14]. The virus consists of a single helical strand of RNA enclosed in a protein sheath 3000 Å long and 180 Å in

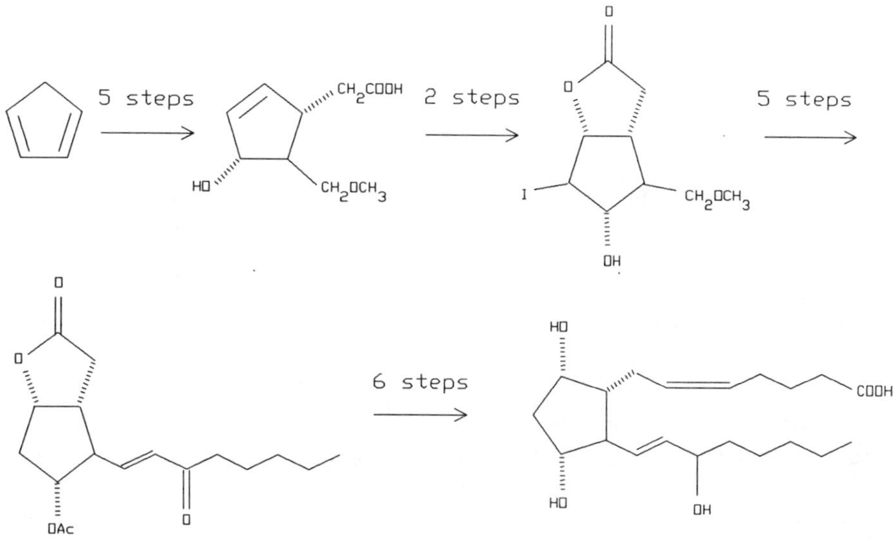

**Scheme 1.1**

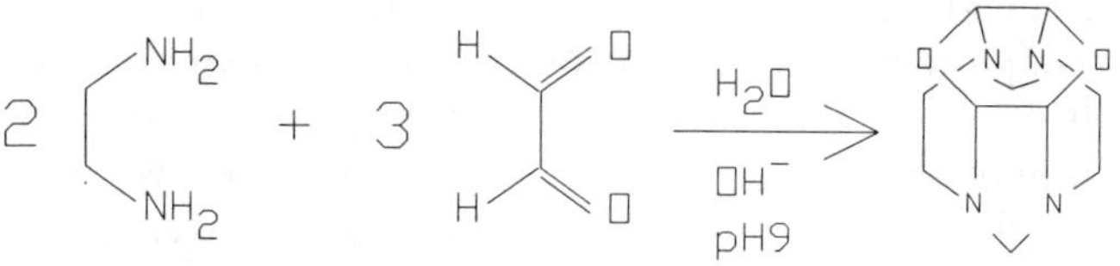

**Scheme 1.2**

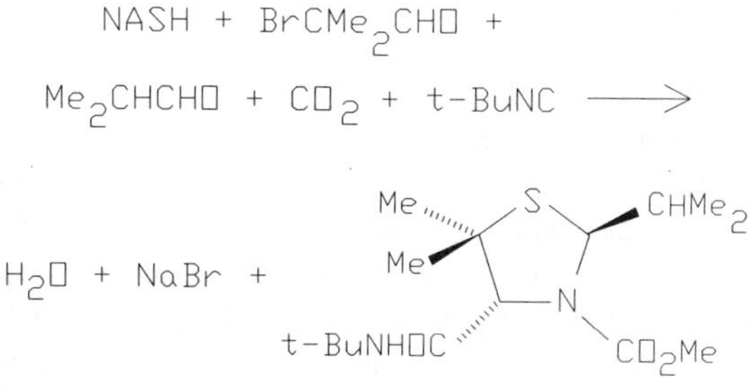

$$NASH + BrCMe_2CH\Box +$$

$$Me_2CHCH\Box + C\Box_2 + t-BuNC \longrightarrow$$

$$H_2\Box + NaBr +$$

**Scheme 1.3**

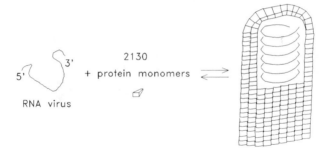

**Figure 1.1**   A schematic presentation of self-assembly of the tobacco mosaic virus. (Adapted from Ref. 12.)

diameter, shown schematically in Fig. 1.1. The sheath is composed of 2130 identical protein monomers. As shown by Fraenkel-Conrat and Williams [10], the virus can be dissociated into its component parts and then reassembled in vitro. What is amazing is the fact that the latter process produces an intact virus. Moreover, the concentration, time, and pH dependence of the reassembling process are characteristic of a typical chemical reaction. The importance of these studies cannot be overestimated. In the nineteenth century Wöhler proved with his synthesis of urea [15] that an organic molecule can be obtained in a chemical reaction from inorganic substrates without involvement of any "vital force." Here, Fraenkel-Conrat and Williams have shown that a living organism can be assembled from molecules each of which can in principle be synthesized in a test tube. It should be stressed that the reconstruction process of the virus is governed by the information contained in the molecules that are its constituent parts. Artificial self-replicating systems make use of the same information-driven processes. One example of such a system is provided by Rebek's reaction presented in Scheme 1.4. There, an addition of the product C to the mixture of substrates A and B surprisingly enhances the reaction yield due to the formation of the intermediate hydrogen-bonded complex D [16a]. Rebek's interpretation of his results was later questioned by Menger and co-workers [16b], showing once more how difficult it is to unravel the secrets of nature.

Biological assemblies are unparalleled in their efficiency in producing error-free constructions on the basis of one or a few repeating units. By self-assembling these units form highly complex structures. Biological assemblies fall outside the scope of this book and will not be discussed here. However, it should be stressed that the understanding of their functioning cannot be achieved without a detailed knowledge of the spatial structure of the assembling components.

Understanding basic stereochemical ideas and their consequences for molecular properties would be impossible without the development of quantum mechanics on one hand and without dramatic progress in physical methods of structure elucidation. With their help many molecules with unusual spatial structures have been synthesized. The structure of some of these molecules will be discussed in this book. Here we would like to mention only few examples of the most startling ones that have been obtained: tetra-t-butyltetrahedrane **4** [17], [1.1.1]propellane **5** [18],

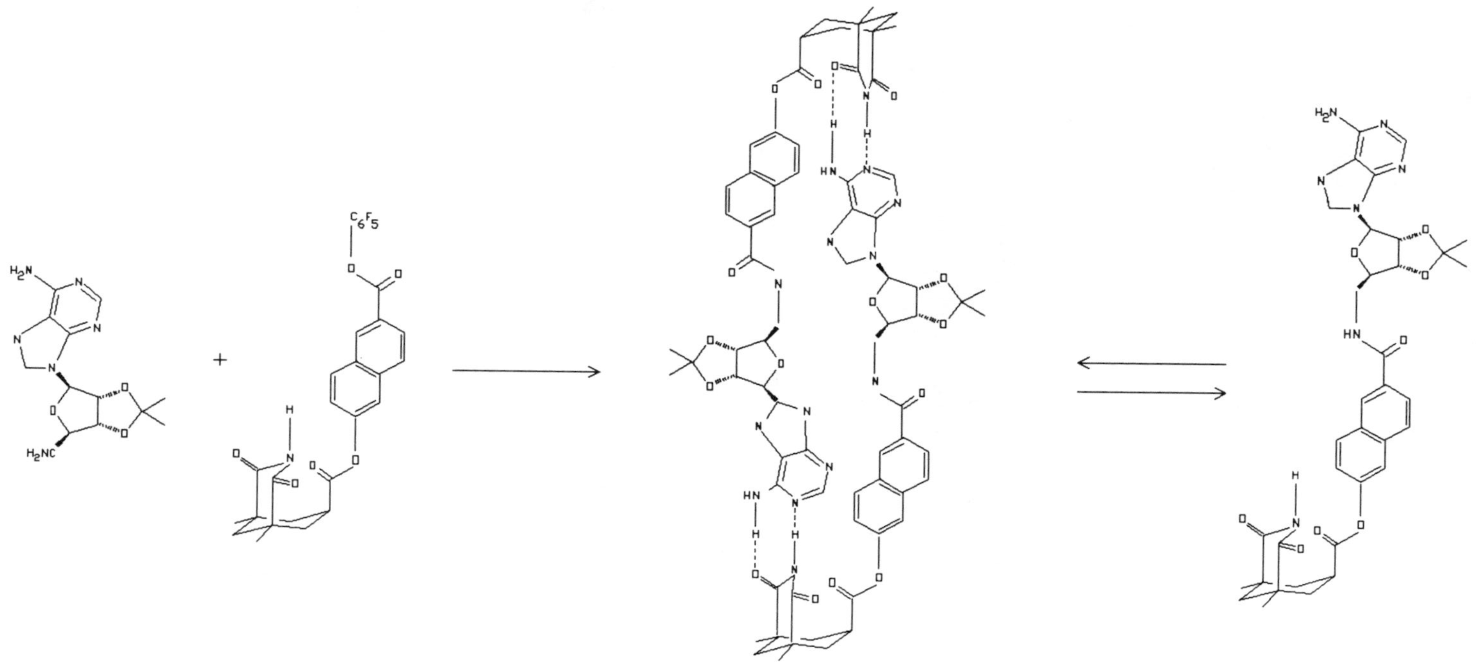

**Scheme 1.4**

A    B    C    D

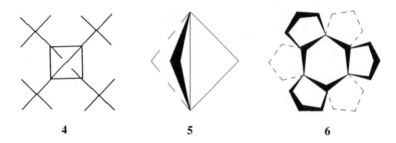

4                    5                         6

[6.5]coronane **6** [19], cubane **7** [20], and dodecahedrane **8** [21], star dendrimers such as **9** [22], [6]-layered cyclophane **10** [23], catenane **12** [24], knot **13** [25], and, last but not least, the elegant fullerene $C_{60}$ **11** [26]. In addition to the beauty of the last structure, numerous practical applications has been proposed for **11** and its derivatives as superconductors, lubricants, drug delivery systems, etc. [27]. It appears almost too good to be true, and in a few years we will learn whether it will stand up to the present high expectations or will serve as another example of an ephemeral mode in science.

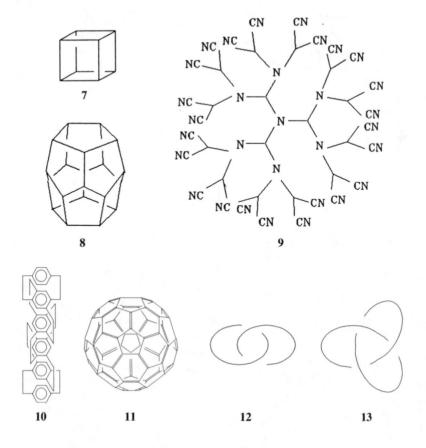

10              11                12                    13

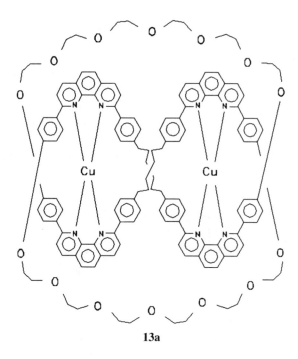

**13a**

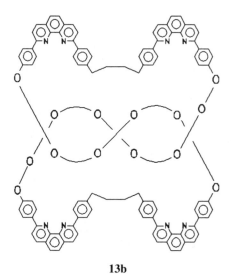

**13b**

The molecules mentioned above are kept together by classical covalent bonding. This is not the case for supramolecular complexes with cyclodextrins (CyDs) **14** [28], crown ethers **16** [29], calixarenes **15** [30], and the beautiful hydrogen-bonded aggregates **17** [31], which are used in diversified practical applications, in addition to raising numerous fundamental questions with their unusual properties. Supramolecular chemistry is a rapidly developing field opening novel perspectives and adding new dimensions to chemistry. At present we do not understand why 1,8-dimethylnaphthalene **18** forms a much stronger complex with β-CyD **14b** than other isomeric dimethylnaphthalenes [32]. We also do not know why the complex of

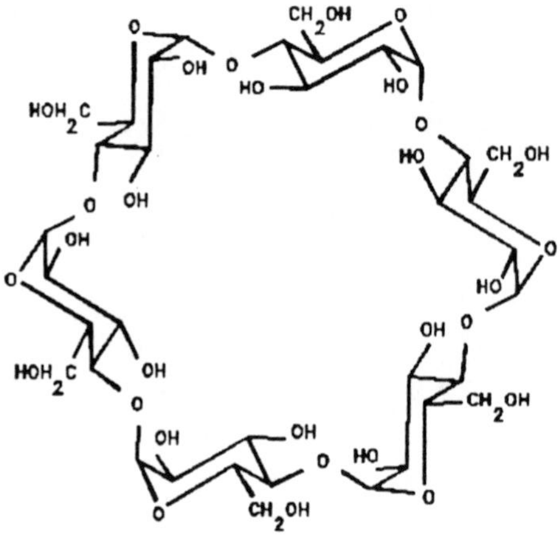

14a

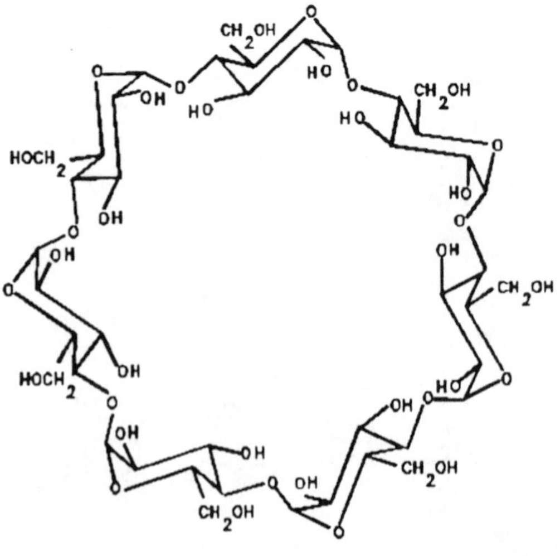

14b

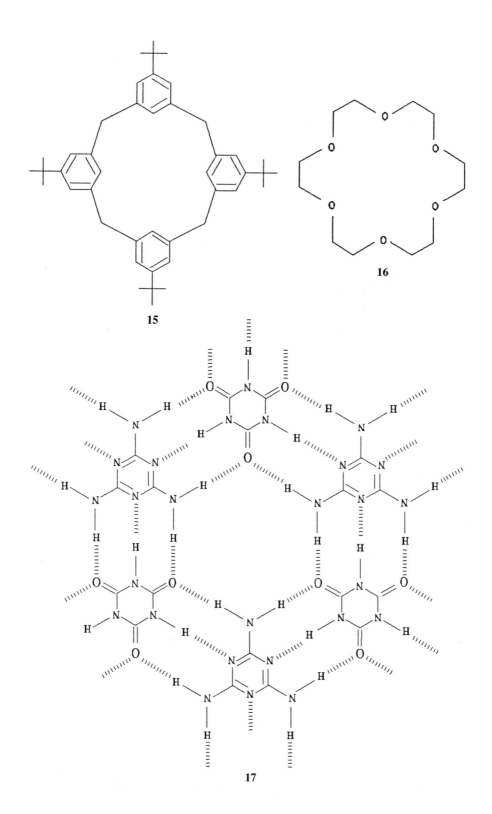

**15**

**16**

**17**

$(-)$-$\alpha$-pinene **19** with $\alpha$-CyD **14a** is stronger than that with its mirror image, $(+)$-isomer **20** (called the enantiomer of **19**; see Chapter 5 on the importance of chirality in nature and in the pharmaceutical industry) [33]. However, it is obvious that the high selectivity and enantioselectivity of the complexation processes critically depend on the complementarity of the shapes of the host and guest molecules. Also in this case, molecular shapes predetermine their properties. These questions will be discussed in Chapter 10, mainly for CyDs.

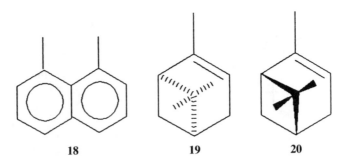

| | | |
|:---:|:---:|:---:|
| **18** | **19** | **20** |

## 1.2  Basic Concepts

A molecule is a dynamic system of atomic nuclei, with their electron clouds executing vibrations around equilibrium positions corresponding to an optimum geometry. According to the Heisenberg uncertainty principle, one cannot precisely determine the positions of nuclei and electrons, which constantly interact with neighboring molecules and radiation. It should also be recalled that at room temperature a real molecule executes rotational and translational movement as a whole. Simultaneously, all its atoms vibrate, and its fragments rotate. The accuracy of modern experimental structure determination techniques is very high, but they use different modes of averaging over the thermal movement of atoms. Thus, as will be shown later, geometrical parameters measured by various techniques differ.

We shall start by treating molecules as a set of point atoms with well-defined positions fixed in space held together by covalent bonds. Such a treatment is certainly an oversimplification and cannot be used for nonrigid molecules such as cyclooctatetraene **21** and other systems discussed in Sections 1.3 and 3.2.2. Moreover, as will be discussed in Chapter 2 for a dynamic system that can exist as a mixture of at least two interconverting forms, the result of its observation depends on the ratio of the rate of exchange between the forms to the time scale of the method. For such dynamic systems one should always specify whether one deals with the averaged structure or individual species. In the case of cyclohexane, these are the planar ring with CH bonds inclined at the same angle to the ring plane **22a** and a usual chair conformation **22b**. There is also a small admixture of the twisted conformation **22c**.

The parameters describing molecular geometry include (Fig. 1.2) the length of a bond formed by atoms $i$ and $j$, $r_{ij}$; the magnitude of the bond angle formed by atoms

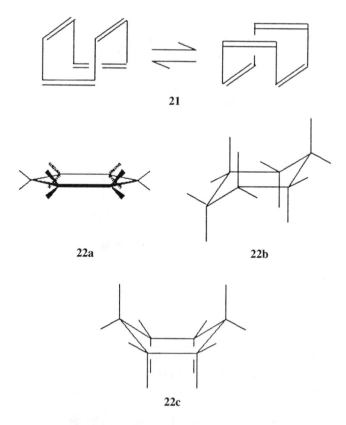

i, j, and k, $\theta_{ijk}$; the magnitude of the torsional angle formed by four, usually not lying in one plane, atoms i, j, k, l, $\Phi_{ijk}$. The sign of the torsional angle $\Phi$ is of importance in many practical problems. The Klyne and Prelog [34] convention has been accepted for its specification. According to this convention, a molecule is oriented in such a way that the plane of the drawing is perpendicular to the central bond

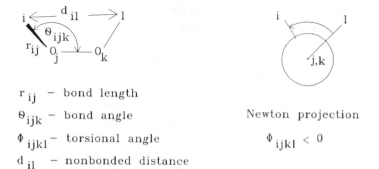

$r_{ij}$ — bond length

$\theta_{ijk}$ — bond angle

$\Phi_{ijkl}$ — torsional angle

$d_{il}$ — nonbonded distance

Newton projection

$\Phi_{ijkl} < 0$

**Figure 1.2** Internal coordinates of the $H_2O_2$ molecule (left) and its Newman projection (right).

of the torsional angle under study. Then the sign of the torsional angle is positive if a movement of the substituent closer to the observer toward the one more distant to the observer is clockwise. Otherwise, as depicted in Fig. 1.2, the sign is negative. In many molecules steric hindrance of substituents that are not covalently bound but lie close in space to each other is of importance. Then nonbonded distances $d_{il}$ between atoms $i$ and $l$ not connected by a covalent bond have to be analyzed.

As mentioned earlier, the values of the geometrical parameters of a molecule under investigation depend on the measurement conditions and experimental techniques used. Such differences are usually of minor significance for synthetic chemists, although together with the charge distribution they determine molecular reactivity. However, their studies become of primary importance when the results of different experimental techniques are compared. They also have to be taken into account when the results of theoretical calculations are compared with experimental data.

Several schematic representations of three-dimensional structure of organic molecules have been proposed [35]. As a rule, only a heavy (nonhydrogen atom) molecular skeleton is given, in which the symbol C for carbon atoms is usually omitted. The bonds lying in the plane of the picture are shown by thin lines of constant breadth. Those pointing out of the paper are denoted by wedges, with the broader end assumed to be closer to the observer (as shown in Fig. 1.2). The ones pointing into the paper are shown by wedges with the thin end away from the observer. Sometimes the bonds situated below the plane of the drawing are shown by dashed lines, or the more distant of two apparently crossing lines is broken to show that they really do not cross in space, as depicted among others in formulae **4, 7,** and **8.**

To show the mutual orientation of atoms around a chemical bond, the so-called Newman projection is frequently used (Fig. 1.2). In this case we look at a molecule along the bond under consideration, with the atom closer to the observer being shown by a circle. The positions of the atoms closer to the observer are given by drawing the bonds connecting these atoms with the central atom closer to the observer. The bonds corresponding to the substituents on the second central atom are seen from behind the first one. Various representation schemes of the spatial structure of molecules have been adopted in different areas of chemistry. The best-known of them are those used in the chemistry of sugars [36]. The Fischer representation of the spatial structure of chain sugars will be shown here. In the Fischer projection the main carbon chain of a sugar is represented by a vertical line with the C1 carbon-bearing carbonyl or carboxyl substituent situated on top of it (see Formulae **1–3**). In such a formula the bonds of the main chain on an asymmetric carbon atom are directed from the observer, while two other substituents on the asymmetric carbon atom are directed toward the observer. The formulae for tartaric acid **1–3** are drawn according to this convention.

Molecular models are very useful tools, enabling a presentation of molecular spatial structure. Space-filling models, sticks (or Dreiding) models, and intermediate ones are the most widely used. In the space-filling models [shown in Fig. 1.3(a) and the color charts] atoms are represented by balls with radii proportional to the

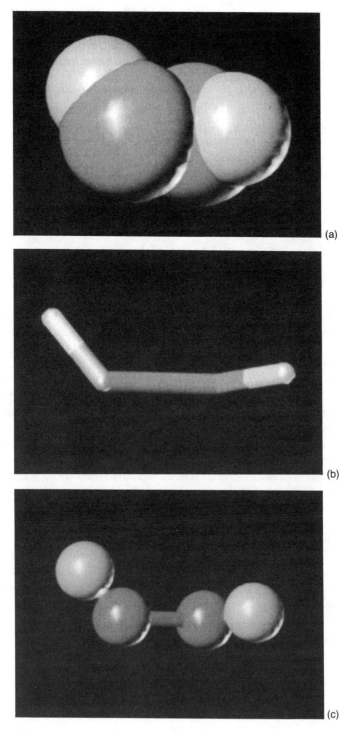

(a)

(b)

(c)

**Figure 1.3**  Molecular models of $H_2O_2$: (a) Space-filling, (b) stick, (c) sticks-and-balls model. (See color plate.)

corresponding van der Waals radii of the atom. The balls are cut off in the directions of bonds formed by the atom at distances proportional to the bond lengths. Contrary to this representation, sticks in the Dreiding models [shown in Fig. 1.3(b) and the color charts] correspond to bonds connected at appropriate angles at a point representing the atom. The first type of these models better shows the molecular shapes, while the second one is more suitable for an estimation of the distances and angles in a molecule. The intermediate models shown in Fig. 1.3(c) try to combine advantages of the former ones. Present computer models frequently replace mechanical toys in chemical laboratories. The computer-generated models shown in Fig. 1.3 and those presented in the color charts were obtained using the program packages SYBYL [37] and KNOW [38].

# 1.3 Reminder of Some Essential Definitions

In 1993 *Chemical Abstracts* registered the twelve millionth chemical compound [39]. Every year the number is growing by more than half a million. Numerous isomeric structures are possible for a given atomic composition. Janoschek reported that for the general formula $C_6H_6$ 217 isomers of this type exist, but only a few of them are known [40]. Fifty years ago the definition of isomers was rather simple. It described substances with the same *summary formula* that could be separated and, roughly speaking, held in different bottles. Nowadays the problem is much more complicated, and experimental conditions have to be specified in which the isomers under study have been observed. Few examples will illustrate this point here (that of cyclooctatetraene **21** was shown in the preceding section, while others will be discussed in Chapter 2 and Section 3.2.2). One type of isomerism is exemplified by the equilibrium between 7-cyano-7-trifluoromethylheptatriene **23a** and 7-cyano-7-methylnorcaradiene **23b**. At $-112°C$ one can observe two forms at a ratio of 78:22 in $^{19}F$ NMR spectrum of the former compound, while at room temperature the averaged spectrum is observed [41]. Molecules **23a** and **b** represent so-called *constitutional isomers*. Another type of dynamic equilibrium is exhibited by chlorocyclohexane. The molecule is known to exist as a mixture of two interconverting chair forms **24a** and **b** with equatorial and axial positions of the substituent. These forms can be separated only at $-150°C$; at room temperature they yield IR spectrum of two forms and NMR spectrum of the averaged one with a planar ring [42]. The NMR manifestation of a dynamic equilibrium involving inversion of cyclohexane rings in *cis*-decalin **25** will be discussed in detail in Section 2.3.7.3.

23a                                    23b

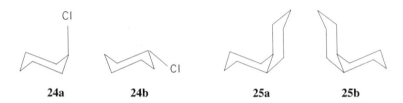

**24a** **24b** **25a** **25b**

Molecules with the same general formula and the same connectivity differing only in the spatial orientation of the constituent atoms are called *stereoisomers*. If two stereoisomers are the mirror images of each other, then such a molecule is chiral, and the stereoisomers are called *enantiomers*. It should be stressed that, without any intervening chiral agent such as another chiral molecule or circularly polarized light, both enantiomers have practically the same chemical and physical properties. The problem of chirality, which is due to the fact that an object cannot be superimposed on its mirror image, will be discussed in more detail in Chapter 5 and Sections 2.2.1.2, 2.3.6, 10.2, 10.3, and 12.1.1.

Stereoisomers that are not enantiomers are called *diastereomers*.

For given experimental conditions the classification of stereoisomers into enantiomers and diastereoisomers is strict (see the discussion of the symmetry and chirality of *cis*-decalin **25** in Chapter 5), since it is based on molecular symmetry. Further division of stereoisomers into conformational and configurational isomers is less precise, since it depends on the activation energy differences between the isomers under consideration. If the difference is larger than 170 kJ/mol (ca. 40 kcal/mol), the stereoisomers can practically be separated, and we are dealing with *configurational isomerism*. This is the case for molecules such as *cis*- and *trans*-decalins **25** and **26,** respectively, (R)- and (S)-chlorobromoiodomethane **27** (see Chapter 5 on chirality for the explanation of the R and S symbols). If, on the other hand, the energy difference is less than 42 kJ/mol (ca. 10 kcal/mol) and stereoisomers cannot be separated, then these are *conformational isomers* or *conformers*. Monosubstituted cyclohexanes, such as **24,** invertomers (i.e., the conformers generated by ring inversion) of *cis*-decalin **25a** and **b,** *trans*- and (+)- and (−)*gauche*-butane **28a–c,** respectively, exemplify this group. In the older literature the term *atropoisomers* was used for stereoisomers with an intermediate energy difference such as some orthosubstituted biphenyls **29.** These stereoisomers can be separated at room temperature, but they are not stable and interconvert.

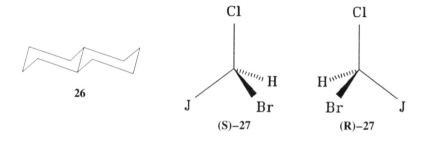

**26**

**(S)–27** **(R)–27**

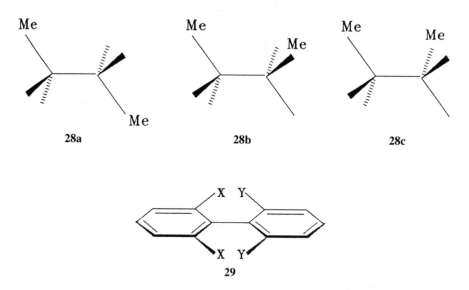

The isomer classification chart (Fig. 1.4) was created by combining the preceding definitions [43]. Conformational enantiomers and diastereomers [such as those presented in Fig. 1.5(a) and (b), respectively] in the scheme call for special

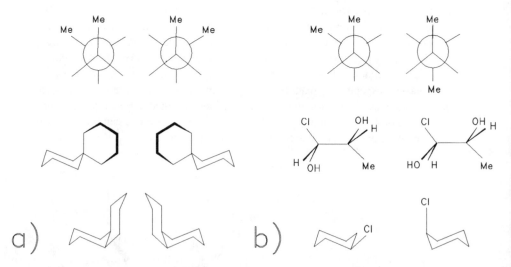

**Figure 1.5** Some examples of (a) conformational and (b) configurational enantiomers.

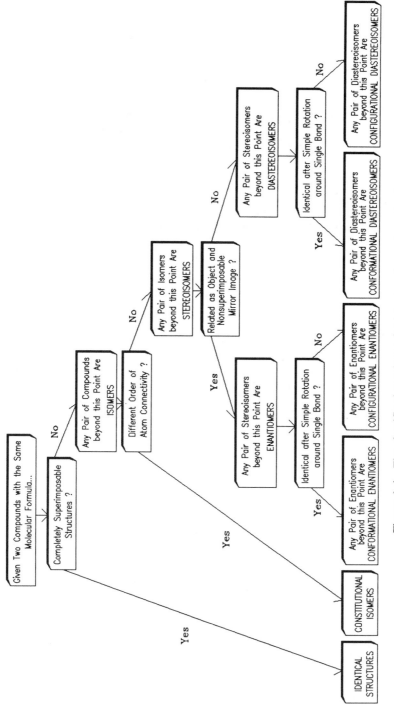

**Figure 1.4** The classification scheme for isomers of organic molecules.

17

attention. As they are nonseparable at room temperature, they are frequently over-looked. However, they play an important role in reaction mechanisms and the operation of drugs. The importance of conformational isomers and diastereomers will be discussed further in Chapter 5.

Several conformations around a single C–C bond are possible. Their definitions are collected in Table 1.1, and some of them are shown in Fig. 1.6. The acronyms "sc" and "ap" in the figure are used for synclinal (that is, close to collinear) and antiperiplanar orientations of the neighboring bonds, respectively. The "+" and "−" signs refer to the sign of the torsional angle.

Topological molecules, discussed in Section 9.1, form a group that can hardly be incorporated into the scheme from Fig. 1.4. They include Walba's ethers modeling a Möbius strip **30** [45], catenanes **12** [24], and knots **13** [25]. Molecules **30** and **13** are chiral; therefore, they exist in form of enantiomers or diastereoisomers. Catenanes **12** and rotaxanes **31** [24] (the latter are not strictly speaking topological molecules;

**Table 1.1**  The Conformer Notation Based on Torsional Angle Values

| Torsional Angle | Klyne and Prelog Notation | Traditional Notation |
|---|---|---|
| $0° \pm 30°$ | $\pm$ syn-periplanar ($\pm$ sp) | eclipsed |
| $60° \pm 30°$ | $+$ syn-clinal ($+$ sc) | $+$ gauche |
| $120° \pm 30°$ | $+$ anti-clinal ($+$ ac) | eclipsed |
| $180° \pm 30°$ | $\pm$ anti-periplanar ($\pm$ ap) | anti, trans |
| $-120° \pm 30°$ | $-$ anti-clinal ($-$ ac) | eclipsed |
| $-60° \pm 30°$ | $-$ syn-clinal ($-$ sc) | $-$ gauche |

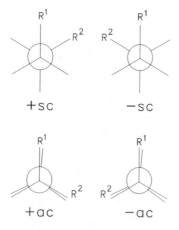

**Figure 1.6**  Some examples of conformations around the C–C bond.

they are included here in view of their similarity to the former ones) are achiral in the absence of additional elements of chirality (see Chapter 5 on chirality). They are isomeric not to a single molecule but to the corresponding pairs of molecules: two rings in case of catenanes, and a ring and a chain molecule for rotaxanes.

As mentioned, a division into configurational and conformational isomers is not well defined, and the observation of conformational isomers critically depends on the method and experimental conditions used. This problem will be discussed in detail in the next chapter, dealing with structure determination by physical methods. Here only a few examples of the energy required to carry out various types of isomerization will be given in Fig. 1.7.

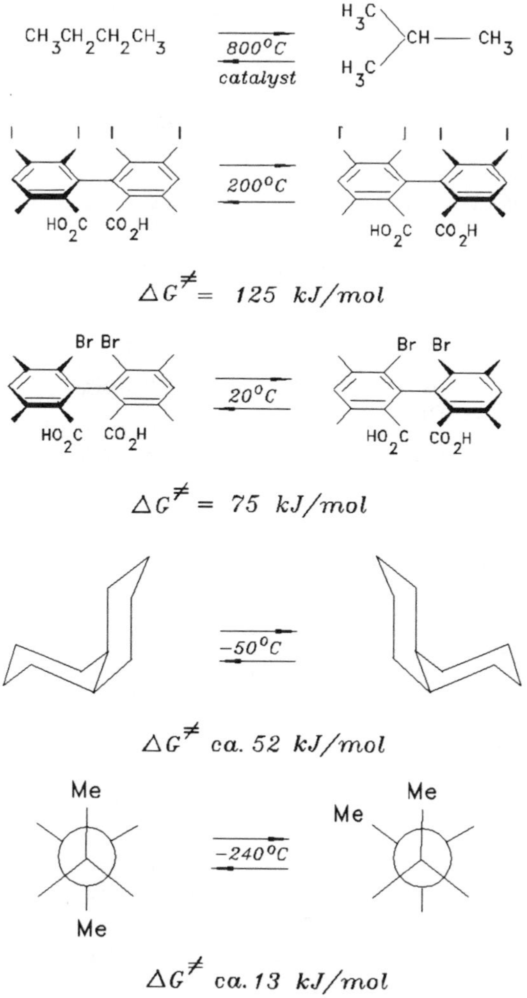

$$\Delta G^{\neq} = 125 \ kJ/mol$$

$$\Delta G^{\neq} = 75 \ kJ/mol$$

$$\Delta G^{\neq} \ ca. \ 52 \ kJ/mol$$

$$\Delta G^{\neq} \ ca. \ 13 \ kJ/mol$$

**Figure 1.7** Some examples of isomeric transformations and the energies associated with them. (Adapted from Ref. 46.)

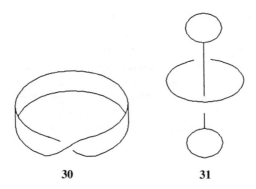

30                          31

## References

1. L. Pasteur, Lectures delivered before the Societe Chimique de France, January 20 and February 3, 1860.

2. J. H. van't Hoff, *Nederl. Sci. Exactes Nat.* (1874) 445.

3. J. A. LeBel, *Bull. Soc. Chim. Fr.* **22** (1874) 337.

4. *Kinderstube* means politeness learned in one's childhood.

5. H. Kolbe, *J. Prakt. Chem.* **15** (1877) 473, cited in F. G. Riddell and M. J. T. Robinson, *Tetrahedron* **30** (1974) 2001.

6. U. Axen, J. E. Pike, and W. P. Schneider, in *The Total Synthesis of Natural Products,* J. ApSimon, Ed., Wiley-Interscience, New York, 1973, p. 81.

7. J. M. Edwards and U. Weiss, *J. Chem. Soc., Chem. Commun.* (1968) 1649.

8. A. Dömling and I. Ugi, *Angew. Chem., Int. Ed. Engl.* **32** (1993) 563.

9. M. Suzuki, A. Yanagisawa, and R. Noyori, *J. Am. Chem. Soc.* **110** (1988) 4718.

10. H. Fraenkel-Conrat and R. C. Williams, *Proc. Natl. Acad. Sci. USA* **41** (1955) 690.

11. K. Namba and G. Stubbs, *Science* **231** (1986) 1401.

12. A. Klug, *Angew. Chem., Int. Ed. Engl.* **22** (1983) 565.

13. C. B. Anfinsen, *Science* **181** (1973) 223.

14. C. J. Epstein, R. F. Goldberger, and C. B. Anfinsen, *Cold Spring Harbor Symp. Quant. Biol.* **28** (1963) 439.

15. Until the end of the eighteenth century living matter was thought to be essentially different from the rest of nature due to a vital force all living creatures contained. This opinion reigned until Wöhler synthesized urea from inorganic molecules.

16a. T. Tjivikua, P. Ballester, and J. Rebek, Jr., *J. Am. Chem. Soc.,* **112** (1990) 1249; J. Rebek, Jr., *Angew. Chem., Int. Ed. Engl.* **29** (1990) 245.

16b. F. M. Menger, A. V. Eliseev, and N. A. Khanjin, *J. Am. Chem. Soc.* **116** (1994) 3613.

17. G. Maier, S. Pfriem, U. Schäfer, and R. Matusch, *Angew. Chem., Int. Ed. Engl.* **17** (1978) 520.

18. K. B. Wiberg and F. H. Walker, *J. Am. Chem. Soc.* **104** (1982) 5239.

19. D. Wehle, N. Schormann, and L. Fitjer, *Chem. Ber.* **121** (1988) 2171.

20. P. E. Eaton and T. W. Cole, Jr., *J. Am. Chem. Soc.* **86** (1964) 962, 3157.

21. R. J. Ternansky, D. W. Balogh, and L. Paquette, *J. Am. Chem. Soc.* **104** (1982) 4503.

22. C. Wörner and R. Mülhaupt, *Angew. Chem., Int. Ed. Engl.* **32** (1993) 1306.

23. T. Otsubo, Z. Tozuka, S. Mizogami, Y. Sakata, and S. Misumi, *Tetrahedron Lett.* (1972) 2927; T. Otsubo, S. Mizogami, I. Otsubo, Z. Tozuka, A. Sakagami, Y. Sakata, and S. Misumi, *Bull. Chem. Soc. Japan* **46** (1973) 3519; T. Otsubo, H. Horita, and S. Misumi, *Synth. Commun.* **6** (1976) 591.

24. G. Schill, *Catenanes, Rotaxanes and Knots,* Academic Press, New York, 1971.

25. C.-O. Dietrich-Buchecker and J.-P. Sauvage, *Angew. Chem., Int. Ed. Engl.* **28** (1989) 189.

26. H. W. Kroto, J. R. Heath, S. C. O'Brien, R. F. Curl, and R. E. Smalley, *Nature (London)* **318** (1985) 162.

27. J. F. Stoddart, *Angew. Chem., Int. Ed. Engl.,* **30** (1991) 70.

28a. *Cyclodextrins and Their Industrial Uses,* D. Duchene, Ed., Editions de Sante, Paris, 1987.

28b. Cyclodextrins are usually abbreviated as CD, which coincides with the acronym denoting circular dichroism. Therefore, throughout this book CyD will be used for these molecules.

29a. J.-M. Lehn, *Angew. Chem., Int. Ed. Engl.* **27** (1988) 89.

29b. F. Vögtle, *Supramolekulare Chemie (Supramolecular Chemistry,* in German), Teubner, Stuttgart, 1989, Chapter 2.

30. C. D. Gutsche, *Calixarenes,* The Royal Society of Chemistry, Cambridge, 1989.

31. C. T. Seto and G. M. Whitesides, *J. Am. Chem. Soc.* **115** (1993) 905.

32. D. Sybilska, M. Asztemborska, A. Bielejewska, J. Kowalczyk, H. Dodziuk, K. Duszczyk, H. Lamparczyk and P. Zarzycki, *Chromatographia* **35** (1993) 637.

33. D. Sybilska and J. Jurczak, *Carbohydr. Res.* **192** (1989) 243.

34. W. Klyne and V. Prelog, *Experientia* **16** (1960) 521.

35. R. Hoffman and P. Laszlo, *Angew. Chem., Int. Ed. Engl.* **30** (1991) 1.

36. *The Carbohydrates, Chemistry and Biochemistry,* W. Pigman and D. Horton, Eds., Academic Press, New York, 1972, Vol. 1A, p. 1.

37. Program package with front-end SYBYL by TRIPOS Associates, 1699 S. Hanley Road, Suite 303, St. Louis, Missouri 63144.

38. K. Nowinski, "Operating instruction to the program package KNOW," Warsaw University, 1990.

39. W. Val Metanomski, *Chem. Intern.,* October, 1993.

40. R. Janoschek, *Angew. Chem., Int. Ed. Engl.* **31** (1992) 290.

41. E. Ciganek, *J. Am. Chem. Soc.* **93** (1971) 2207; E. L. Eliel, *Isr. J. Chem.* **15** (1976/77) 8.

42. F. R. Jensen and C. H. Bushweller, *J. Am. Chem. Soc.* **88** (1966) 4279.

43. The scheme was developed by H. Dodziuk, *Tetrahedron. Asym.* **3** (1992) 43, by modifying the erroneous Black scheme [44].

44. K. A. Black, *J. Chem. Ed.* **67** (1990) 141.

45. D. M. Walba, *Tetrahedron* **41** (1983) 3161.

46. M. Nogradi, *Stereochemistry, Basic Concepts and Applications,* Academiai Kiado, 1981, p. 38.

# *2*

# Physical Methods as a Source of Information on the Spatial Structure of Organic Molecules

## 2.1 Introduction

Fifty years ago, to get information on the composition of a newly synthesized substance, an organic chemist had to carry out laborious degradative reactions accompanied by elemental analyses. This involved weighing, dissolving, allowing reactions, isolating products, purifying and the analyzing. A subsequent synthesis of a natural compound having the same properties as the one under study served as proof of its structure. Such tedious procedures enabled one to go from a *summary* formula $C_4ONH_9$ to $CH_3CON(CH_3)_2$ on the basis of the following reasoning: (1) six H atoms behave in a different way than the other three; (2) two carbon atoms behave in a different way than the third one. This kind of logic checked over thousands of compounds proved remarkably successful, allowing one to transfer information from chemical phenomena into the formulas, forming the foundation of chemistry. By means of physicochemical methods, the formulas were later shown to represent the structure of organic molecules. Establishing the composition of a molecule by a chemical degradative procedure could easily take many years, and the substance was destroyed in course of the study. Today the use of new, noninvasive physicochemical techniques and the application of stereochemical concepts provide much more detailed information, not only about the composition, but also about the spatial structure of a medium-sized organic molecule in minutes, hours, or, in more complicated cases, days without destroying it.

Establishing the structure of patchouli alcohol **32b** demonstrates the difference between chemical and physical approaches and their effectiveness [1a, 2]. A long series of chemical experiments aimed at the structure determination of this sesquiterpenoid compound lasted more than 50 years and was considered to be completed by its total synthesis [3]. The result was structure **32a**, while the X-ray study in early 1960s that took a few weeks unambiguously yielded **32b** as the structure of patchouli alcohol.

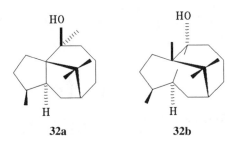

**32a**                         **32b**

The aim of this chapter is to show briefly the most important experimental methods that allow one to obtain reliable information concerning molecules as three-dimensional objects, that is, their bond lengths, bond angles, and torsional angles. Much less emphasis will be placed on another important aspect, namely, on molecular energy. Due to space limitations, the foundation of each method will be presented briefly, followed by a few interesting examples of its applications. As mentioned earlier, a molecule is not a static system: It moves as a whole, its constituent atoms vibrate, and its fragments rotate, executing thermal motion. Various methods of structure determination have different time scales of observation (summarized in Table 2.1 [4]) that enable one to observe separate signals for different conformers. The methods are based on different physical phenomena, and different modes of averaging over this motion are applied in them. For this reason they yield slightly different values of bond lengths and angles. This prompted us to compare the accuracy of information on geometrical parameters of a molecule stemming from different methods, with the emphasis laid on their complementarity. Two main groups of physicochemical methods will be discussed, namely, *diffraction* methods and *spectroscopic* ones. In addition, a new method of a different kind, *scanning probe microscopy,* will be briefly presented at the end of this chapter. In both diffraction and spectroscopic methods a sample of the substance to be analyzed or of its solution is subjected to electromagnetic radiation having a wavelength $\lambda$ corresponding to the desired process. For instance, absorption of radiation with a

**Table 2.1**   Time Scale of Some Physicochemical Methods
of Structure Determination (in $s^{-1}$)

| Method | Time scale | Method | Time scale |
|---|---|---|---|
| Electron diffraction | $10^{-20}$ | Visible spectroscopy | $10^{-14}$ |
| X-ray diffraction | $10^{-18}$ | IR Raman spectroscopy | $10^{-14}$ |
| Neutron diffraction | $10^{-18}$ | MW spectroscopy | $10^{-10}$ |
| UV spectroscopy | $10^{-15}$ | NMR spectroscopy | $10^{-1}$–$10^{-9}$ |

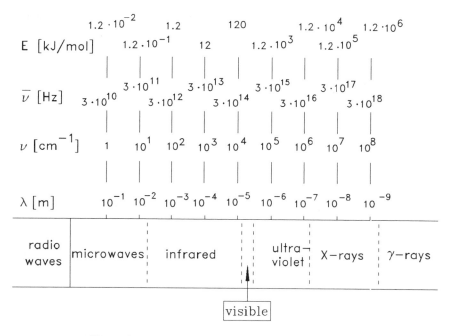

**Figure 2.1**   The spectrum of electromagnetic radiation.

wavelength λ of ca. $10^{-7}$ m causes electronic transitions, that with λ ca. $10^{-5}$ m is related to molecular vibrations, radio waves with λ greater than $10^{-1}$ m bring about transitions between various nuclear spin states in NMR spectra, etc. A division of the electromagnetic spectrum into regions is presented in Fig. 2.1. From this scheme one can see that almost the whole range of radiation is used in structural studies. It encompasses radio waves with frequencies less than ca. $3 \times 10^{10}$ Hz (and a corresponding wavelength greater than ca. 1 $cm^{-1}$), which are applied in *nuclear magnetic resonance* (NMR); the *microwave* (MW) region, in which absorption is associated with rotations of molecules or their fragments; the *infrared* (IR) region, in which molecular vibrations are studied; the *ultraviolet* and *visible* (UV-VIS) regions, in which electronic transitions are analyzed; and X radiation.

## 2.2 Diffraction Methods

Diffraction methods encompassing X-ray, neutron (ND), and electron diffraction (ED) provide the most detailed information on the spatial structure of molecules. In the structure elucidation of organic molecules, the first two are usually applied in studies of crystal structures, while the third yields information on structure in the gas phase. At present, X-ray analysis and neutron diffraction allow one to study crystals smaller than $1 \times 10^{-4}$ $cm^3$. The necessity of having crystals limits the applicability of these methods on one hand. On the other hand, it has forced the creation of a new, rapidly expanding field dealing with crystal growth. In particular, obtaining crystals of peptides has recently enabled structural studies of these molecules by diffraction methods. This in turn has opened new horizons in biochemistry.

A concise introduction to crystalline structure is necessary [5]. In the crystalline state atoms, ions, and molecules (or parts of molecules or groups of molecules) are arranged in ordered, repeating patterns (they are said to have *translational symmetry*), and the smallest repeating part is called a *primitive elementary cell*. The cell is characterized by the edge sizes *a,b,c* and by the angles they form. There are only 14 types of crystal lattice systems, as schematically presented in Fig. 2.2. X rays, neutrons, and electrons are diffracted by electrons or nuclei.

For a given crystal various crystallographic planes can be distinguished. Let us look at the different planes in a two-dimensional lattice as presented in Fig. 2.3. Neglecting the unit cell sizes *a* and *b,* these lattices can be characterized by three

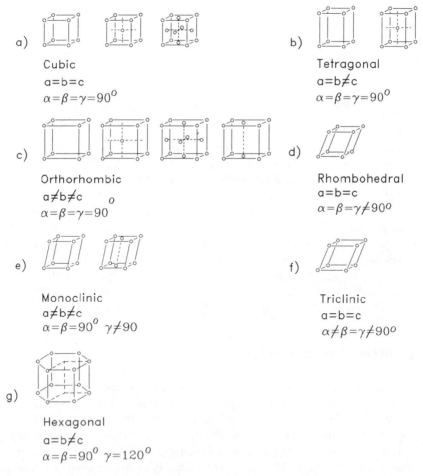

a)
Cubic
a=b=c
$\alpha=\beta=\gamma=90^{o}$

b)
Tetragonal
a=b≠c
$\alpha=\beta=\gamma=90^{o}$

c)
Orthorhombic
a≠b≠c
$\alpha=\beta=\gamma=90^{o}$

d)
Rhombohedral
a=b=c
$\alpha=\beta=\gamma\neq90^{o}$

e)
Monoclinic
a≠b≠c
$\alpha=\beta=90^{o}$  $\gamma\neq90$

f)
Triclinic
a=b=c
$\alpha\neq\beta=\gamma\neq90^{o}$

g)
Hexagonal
a=b≠c
$\alpha=\beta=90^{o}$  $\gamma=120^{o}$

**Figure 2.2**   The fourteen Bravais crystal lattices.

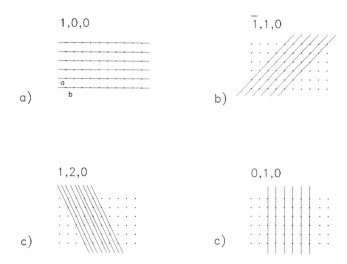

**Figure 2.3** Some of the crystallographic planes that can be drawn through the space lattice and their corresponding Miller indices (*hkl*). A bar over the index denotes its negative value.

numbers (the last one, zero, corresponds to the third dimension, omitted in the two-dimensional case). Reciprocals of these numbers, called the Miller indices *hkl*, define the lattice reciprocal of the one investigated. They are very practical for expressing the separation of crystallographic planes. For instance, for a general orthorhombic lattice [Fig. 2.2(c)] built from rectangular elementary units with three different sides *a,b,c,* the distances between the *hkl* planes are given by the formula

$$1/d^2_{hkl} = (h^2/a^2) + (k^2/b^2) + (l^2/c^2) \qquad (2.1)$$

## 2.2.1 X-Ray Crystallography

### 2.2.1.1 The X-Ray Method, Its Merits and Limitations

In the X-ray method of structure determination [1], the diffraction of X rays (with a wavelength $\lambda$ varying between 0.1 Å ($10^{-11}$ m) and 100 Å ($10^{-8}$ m) is performed by electrons of atoms or ions situated at crystallographic positions. As a first approximation, the atoms in a molecule can be treated as spherical objects. With this approximation relatively small effects related to chemical bonding (for atoms different from the hydrogen atom) and more significant effects associated with the anisotropy of thermal motion are not taken into account. If a monochromatic wave falls on a crystal, then it is diffracted (Fig. 2.4). As shown by Bragg, the diffraction process can be described as interference of reflections from crystallographic planes. The wavelets "reflected" by various crystallographic planes are reinforced if the difference in distances traveled by these waves corresponds to an even multiplicity of the wavelength. This condition is known as the *Bragg equation*

$$n\lambda = 2d \sin \theta \qquad (2.2)$$

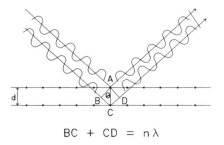

$$BC + CD = n\lambda$$

**Figure 2.4**   Reflected in-phase X-ray waves.

where $n$ is an integer, $\lambda$ denotes the wavelength of the incident and scattered waves, $d$ is a distance between crystallographic planes (shown in Fig. 2.3 in the simplest case of a square planar lattice if $a = b$) and $\theta$ is the angle of incidence of the wave. On the other hand, if the difference in the distances between the waves scattered by atoms situated at two crystallographic planes ($BC + CD$) is equal to an odd multiplicity of half of the wavelength, then the waves cancel each other. As a result of the reinforcements and cancellations, a diffraction pattern is produced, for example, on a photographic plate. Dark spots on the plate called *reflections* are related to different $n$ values and different distances $d$ corresponding to various interatomic distances in the crystal under investigation. Of course, the situation in a three-dimensional crystal requires a more complicated analysis to determine bond lengths and angles. However, the diffraction pattern allows one to determine quickly the type and symmetry of the crystal lattice. The elucidation of bond lengths and angles from the pattern is a more difficult task. The observed intensity of reflections obtained by the scattering $I_{hkl}$ are proportional to $|F_{hkl}|^2$, where the *structure factor* $F_{hkl}$ is the ratio of the *scattering amplitude* by the contents of an elementary cell to the scattering amplitude of a single electron situated at the center of the coordinate system of the cell. $h$, $k$, and $l$ denote the Miller indices of the crystal lattice under study.

$$F_{hkl} = \sum_{\text{atoms } j}^{N} f_j \, e^{-2\pi i(hx_j + ky_j + lz_j)} \tag{2.3}$$

in the elementary cell

where $f_j$ are *atomic scattering factors*. Structure factors are related to the electron density distribution in an elementary cell $\rho(r)$

$$F_{hkl} = \int \rho(r) \, e^{-2\pi i(hx + ky + lz)} \, dV \tag{2.4}$$

It is very important that, in addition to the coordinate dependence exhibited by the preceding formula, $F_{hkl}$ contains temperature factors that are usually anisotropic. This enables one to study the thermal motion of atoms and molecules in the crystal. To obtain bond lengths and angles of the molecule under investigation, one has to assume certain values for these parameters. Then the structure factors $F_{hkl}$ and phase factors $\phi_{hkl} = \exp[-2\pi i(hx+ky+lz)]$ are calculated for the assumed structure. After calculating the values of structure factors and phases, an iterative procedure is carried out consisting of changing the geometrical parameters in such a way that the best reproduction of the experimental diffraction pattern is obtained. It should be stressed that the intensity of the reflections $I_{hkl}$ is proportional to the sum of the structure factors squared. Therefore, the determination of the signs of the phase factors is an ill-defined task, complicating the structure elucidation. The accuracy of the results of an X-ray analysis depends on the number of reflections observed. The *residual index R* is a measure of the accuracy

$$R = \frac{\Sigma_h[|F_o(h)| - |F_c(h)|]}{\Sigma_h|F_o(h)|} \tag{2.5}$$

where the structure factors with the subscripts o and c denote the values observed and calculated on the basis of the assumed values of the geometrical parameters, respectively.

### 2.2.1.2 Determination of Absolute Configuration

In the model briefly sketched above, so-called normal scattering was assumed. This means that the phase differences between waves scattered at different points depend only on path differences. This, in turn, implies that any intrinsic phase change connected with the actual scattering must be the same for all scattering centers. This assumption is usually quite a good approximation. If not, then there is a slight difference between the intensities of the $(hkl)$ and $(-h,-k,-l)$ (denoted as $\bar{h},\bar{k},\bar{l}$ ) reflections, enabling one to establish absolute configuration of chiral molecules. This method was proposed by Bijvoet [6]. Until the first determinations of absolute configuration by this method in the early 1950s, only relative configurations between similar molecules or different asymmetric centers in the same molecule could be established. Thus the validity of Fischer's assumption on the absolute configuration of glyceraldehyde **33** could be confirmed only after 60 years [7]. Another study [8] confirmed the correctness of the formula **34** for the most common L configuration of natural amino acids [9].

**33**                                                    **34**

### 2.2.1.3 Structural Correlation Principle

The study of potential energy hypersurfaces showing their dependence on bond lengths and angles values etc., is indispensable for the understanding of chemical reactivity. The surfaces can be studied theoretically only for very small molecules, but some information on their structure can be obtained by searching for certain types of correlations among the structural parameters for a related group of molecules or crystals. The following procedure is adopted for this purpose [1b, 10]. X-ray data for molecules possessing a structural fragment or subunit under study (later called sample points) are chosen in such a way as to reflect the diversity in its environment. Then one looks for correlations among independent geometrical parameters of the fragment. The structural parameters in every molecule correspond to the equilibrium values under the perturbation operative in this structure. Thus, in addition to the scatter of the parameter values due to experimental inaccuracies, some spread of their values must result from the different perturbations affecting the fragment under study in its different environments. The effect of these perturbations is thought to be averaged on the basis of a further assumption that irregularities in the distribution of sample points are not too large. Then the averaging should yield a more or less well-smoothed curve approximately following a minimum energy path in the parameter space. The latter means that the fragment or subsystem deforms along the path of least resistance.

The energy of the system depends on a large number of geometrical parameters, and the choice of effective ones is a difficult task, similar to the choice of reaction coordinates in quantum calculations. This will not be discussed here in detail. The interested reader can find a detailed presentation of the structural correlation principle and its applications in Dunitz's book [1b] and in Ref. 10b.

### 2.2.1.4 Summary

At present, X-ray analysis is the method most frequently used in structural studies of molecules in the solid state. It provided a solid basis for studying the spatial structure of molecules. The results obtained are collected in crystallographic databases. With over 100,000 files, the 1993 Cambridge Structural Database [11a] contains the most extensive set of such data for organic and metallorganic molecules, while the corresponding, much less extensive data for peptides [11b] are collected in the Brookhaven Database. The processing of the massive amount of data collected in these databases is carried out by computers [11c]. An interesting example of the valuable information extracted from such an analysis was recently reported by Gilli and co-workers [11d]. The authors evaluated X-ray and ND results on O—H...O hydrogen-bonded systems with strong and very strong H bonds. The study of structural correlations enabled the authors to classify the bonds under investigation into three principal types: (a) negative charge assisted, (b) positive charge assisted, and (c) resonance-assisted H bonds. In the third case, the two oxygen atoms are interconnected by a system of $\pi$-conjugated double bonds.

The limitations of X-ray structural analysis consist of (a) the necessity of having a monocrystal of the molecule under study and (b) the fact that the results obtained do not refer to an isolated molecule but describe the one on which crystal forces are imposed. These forces usually do not affect the X-ray results significantly. However, molecules for which several conformers exist in the gas phase or in solution as a rule crystallize in one, most frequently the most stable, conformer. The differences between the conformation in the gas phase (or liquid) and that in the crystal arises in the case of rotational barriers less than or comparable to the magnitude of the crystal forces. The best-known example of this kind is biphenyl **35**. In the gas phase its phenyl rings are inclined at an angle of ca. 45°, while in the most stable crystal modification the molecule is planar [12]. On the other hand, in some cases so-called conformational polymorphism is possible; that is, molecules adopt different conformations in different crystal modifications [13]. An interesting example of different solid-state and solution structures is provided by a cyclodextrin **36a** complex with *para*-nitrophenol **37c** ($X = NO_2$), discussed in Section 2.4.2 [14]. Other examples demonstrating the insensitivity of the X-ray method to the mobility of a guest molecule in cyclodextrin complexes will be discussed in Sections 2.3.7 and 10.3.

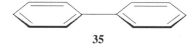

**35**

A considerable disadvantage of X-ray analysis is the smaller accuracy in the determination of hydrogen atom positions (see Section 2.2.2 and Ref. 15 for a comparison of the accuracy of X-ray and ND determinations of hydrogen atom positions). In general, the accuracy in the determination of bond lengths and angles depends on the magnitude of the molecule under investigation. By very precise determinations it is possible to study the thermal motion of atoms in a crystal.

Thermal motion can also obscure X-ray results. In addition to the random motion in crystals, there is the possibility of collective motion of molecular fragments even in the solid state. Thus methyl groups in hexamethylbenzene **38** [16], hydroxyl groups in β-cyclodextrin **14b** [17], or perchlorate groups $ClO_4^-$ [18] rotate at room temperature in a crystal. Studying such motions can provide valuable information on molecular dynamics in the solid state. However, they can also restrict the possibility of a proper interpretation of X-ray data. An interesting example of this kind is provided by the study of $C_{60}$ **11**. As discussed in Section 2.3.4, the spherical shape of this molecule allows for its considerable mobility in crystals at room temperature, prohibiting the determination of the bond lengths. Under these conditions they have been determined first using $^{13}C$ NMR [19].

The independence of structural parameters on isotopic substitution is assumed in most structural studies. This assumption holds well for atoms other than hydrogen, but it can be too crude an approximation for light hydrogen atoms undergoing

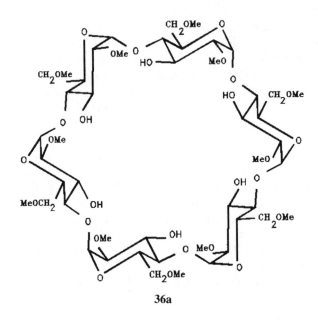

36a

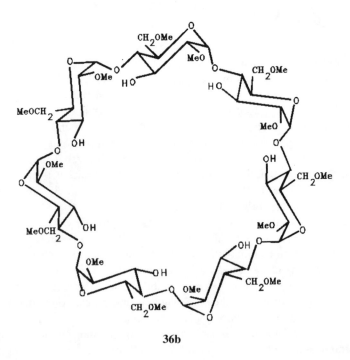

36b

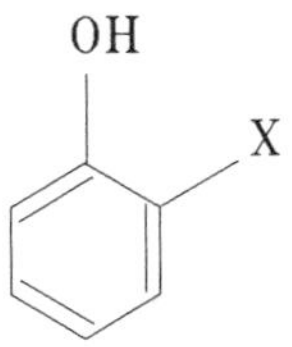

37a 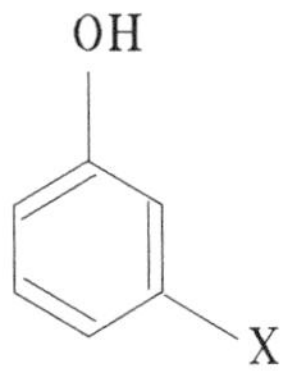          37b 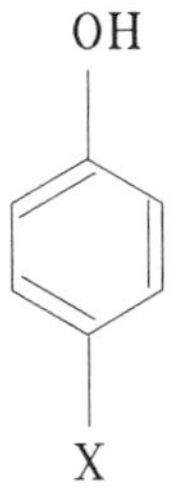          37c 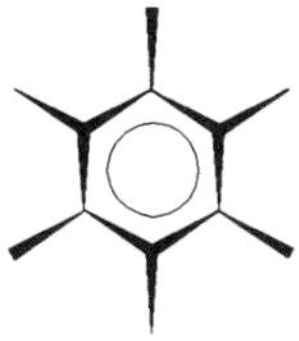          38

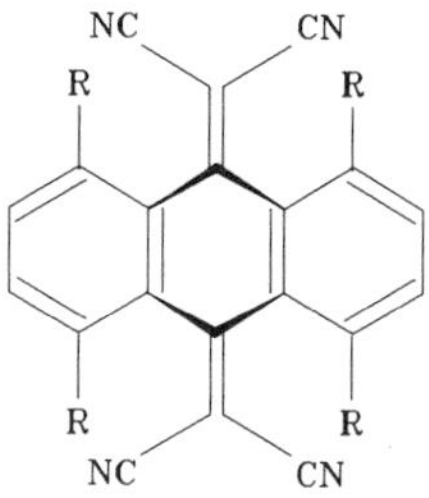

R = H, D

39

considerable anharmonic motion. An interesting example of this kind has recently been published [20]. The X-ray structure of tetracyanoanthraquinodimethane **39** exhibits a nonplanar distortion that decreases considerably after deuteration. The effect is due to the smaller anharmonicity of the CD stretching vibrations than that of the CH bond, causing a decrease in the effective lengths of CD bonds.

The distances measured by X-ray analysis refer to the values for atomic electron densities averaged over thermal motion. Therefore, they differ slightly from the values corresponding to the internuclear distances determined by other techniques. The difference is significant for the determination of CH bond lengths since the electron density is shifted from the H atom to carbon in the CH bond.

## 2.2.2 Neutron Diffraction

*Neutron diffraction* method [21] (ND) is similar to X-ray structure determination, but there are two essential differences between them. First, in ND a sample is irradiated with a slowed beam of neutrons produced in a nuclear reactor instead of X rays. For this reason the method is more expensive than X-ray analysis. Second, the diffracting medium in this method is atomic nuclei, not the valence electrons, as is the case with X-ray analysis. Therefore, the diffraction processes differ, and the information obtained by means of ND is complementary to that gained by X-ray analysis. In particular, hydrogen atom positions are more precisely determined, and the thermal motion of atoms in the molecule under investigation can be more accurately studied.

Typically, the neutrons used correspond to a wavelength 1.8 Å (180 pm). At room temperature they have an energy of 0.025 eV. Thus, similarly to the situation in X-ray analysis, the neutron wavelengths are equal to typical interatomic distances in crystals, but their energy is comparable to the thermal energy of atomic vibrations in crystals. This also constitutes a significant difference between X-ray analysis and neutron diffraction, since in the former method the photon energy is of the order of $10^3$ eV. Therefore, inelastic scattering, which, to a good approximation, can be neglected in X-ray analysis, is an important factor in ND, allowing one to analyze precisely the thermal motion of nuclei. Both techniques are based on diffraction, obeying the Bragg equation, and the intensity observed in ND is also proportional to the sum of the squares of the structure factors $F$. Thus, similarly to the X-ray analysis after solving the phase problem, one calculates structure factors $F_{hkl}$. Then the assumed geometrical parameters are modified to reach the best agreement of the calculated diffraction pattern with the experimental one. The same mathematical procedures are used to determine geometrical parameters by means of neutron diffraction and X-ray analysis.

ND can be applied to the study of magnetic structures, liquid, and gaseous species and to examine inelastic scattering, etc. However, the most important application of the method in the structural analysis of organic molecules consists in an accurate determination of the positions of light atoms (in particular, hydrogen) in the solid state by means of the elastic scattering of neutrons, since the hydrogen atom positions are more precisely determined by ND than by X-ray analysis, as illustrated by studies of the H-bonding pattern in β-cyclodextrin **14b** carried out by Saenger and co-workers [17]. Hydrogen atom positions in this molecule could not be determined by X-ray analysis unequivocally, and it was not possible to elucidate its hydrogen bonding pattern. ND studies by Saenger and co-workers at 293 and 120 K have shown that the O2H2 and O3H3 groups in **14b** can be engaged in hydrogen bonds in two different ways: Namely, an H atom attached to O2 can form a bond with the neighboring O3 atom, or, alternatively, the O3H3 bond can point toward the neighboring O2 atom, with the H3 hydrogen atom forming an H bond with it. At room temperature these hydrogen atoms are distributed over 16 sites with ca. 0.5 occupancy [17b]. A low-temperature neutron study [17a] has proved that the disorder established in the X-ray study is not the statistical one. This means that at room temperature the circular patterns of H bonds pointing in opposite directions (H2→O3 or H3→O2) rapidly interchange in concerted thermal motion, even in crystals. Most of the 293 and 120 K ND data were reproduced by molecular dynamics simulations by Koehler et al. [17c], and model semiempirical quantum calculations [17d] showed that a cooperative system of H bonds forming a circle is energetically favored.

The possibility of analyzing precisely the position of hydrogen atoms (including positions of hydrogen atoms of crystalline water) is especially important in peptide studies, since it allows one to determine the hydrogen bonding pattern defining their spatial structure.

## 2.2.3 Electron Diffraction

*Electron diffraction* (ED) [22] makes use of the wave character of electrons, in analogy to neutron diffraction, which makes use of the wave character of neutrons. Contrary to X-ray and ND, mainly gas-phase species are studied by this technique. The differences in the ED pattern in comparison to the former methods are due to differences in the diffracting factors and those due to the different state of the molecule under study. In ED we deal with electron scattering by molecules executing thermal motion, thus changing their mutual random orientation. Therefore, the observed quantities have to be averaged over all possible orientations of the diffracting molecules. Thus, in ED, the scattering intensity $sM(s)$ is studied as a function of the $s$ value, which is equal to

$$s = [4\pi \sin(\vartheta/2)]/\lambda \tag{2.6}$$

where $\vartheta$ is the angle between incident and scattered beams and $\lambda$ is the wavelength of the scattered radiation. Similarly to X-ray and ND analysis, the peak intensity is a function of the structure factors $F$, but the functional dependence relating these quantities is different.

Sometimes another function that is usually used in ED, namely, the radial distribution function $f(r)$, which exhibits maxima corresponding to the interatomic distances in a molecule under study, is given. The typical dependence of $f(r)$ on $r$ is given in Fig. 2.5 (curve E) [23] for 1,2-ethan-dithiol **40**. Such curves allow one not only to determine the bond lengths but also to obtain the distances between atoms that are not chemically bonded. Once more, to interpret the ED observations, one has to assume the geometrical parameters of the molecule under investigation. Then geometrical parameters are fitted by a trial-and-error procedure to obtain the best possible agreement between the experimental and calculated curves. The difference between these curves, showing the accuracy of the fit, is plotted in the bottom curve of Fig. 2.5. In many cases, some peaks in the radial distribution curve can be interpreted in different ways. To remove this ambiguity in ED data, some additional information is necessary. For instance, to assign the peaks corresponding to C—C and C—N distances in azetidine **41** [24], one can make use of the general information that all known C—C bonds are longer than the C—N ones. Ascribing the corresponding peaks to C···C and C···N diagonals in the four-membered ring is a much more difficult task. It cannot be solved on the basis of ED experiments alone without the involvement of other physicochemical methods or theoretical considerations.

HS(H$_2$C) —— (CH$_2$)SH

**40**                                                                          **41**

In favorable cases ED allows one to determine the conformational equilibrium in a molecule under investigation. This application is also illustrated in Fig. 2.5 [23]

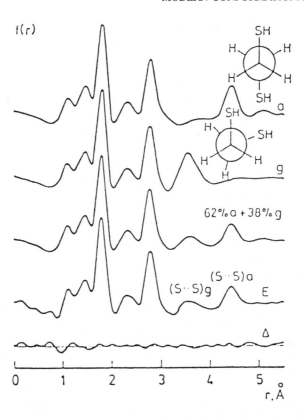

**Figure 2.5**    The experimental and calculated radial distribution curves for ethan-1,2-dithiol **40.** *a* and *g* denote the calculated curves for the *trans-* (i.e., anti-) and *gauche-*conformers. As described in the figure, the next curve corresponds to the calculated values for a mixture of 62% of the former and 38% of the latter conformers, respectively. The experimental curve revealing separate distinct peaks for S···S nonbonding distances of the *trans-* and *gauche-*conformers is denoted by E, while the bottom curve represents the difference between the experimental and best-fitted calculated curves. (Courtesy of Prof. I. Hargittai.)

for ethan-1,2-dithiol **40.** The calculated curves for the pure *trans-* (called anti in [23]) and *gauche* forms [the curves denoted by (a) and (g), respectively, are also given in Fig. 2.5, together with the calculated curve for the mixture consisting of 62% of the *trans* form and of 38% of the *gauche.*] As described earlier, the lowest curve in Fig. 2.5 corresponding to the difference between the best calculated and experimental ones shows the accuracy of the ED determination of the geometrical parameters of **40.** A study of the temperature dependence of the equilibrium constant between these forms can, in favorable cases, yield information on the thermodynamic parameters of the investigated molecule. Analogous results have been obtained, among others, for cyclodecane $C_{10}H_{20}$ [25]. As discussed in Section 3.3, to interpret experimental data in this case, the use of model molecular mechanics calculations was indispensable.

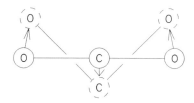

**Figure 2.6** Bending of the $CO_2$ molecule, resulting in the "shrinkage effect."

The bond lengths and angles determined by means of ED refer to the averaged values. Therefore, these values are not consistent and reveal the so-called "shrinkage effect." Its manifestation are most easily seen for linear molecules for which the distance between nonbonded atoms determined by this technique is less than the sum of the corresponding bond lengths. Figure 2.6 illustrates this point for $CO_2$ molecule [26]. Here, owing to the bending vibrations of the OCO angle, the mean value of the distance between the oxygen atoms is lower than the mean value of the sum of the CO bond lengths. Similarly, owing to the *shrinkage effect*, linear molecules exhibit apparent nonlinearity. For example, the value of the C≡C—H angle in acetylene **42** determined by ED is equal to 7.8° [27].

$$H \!-\! C \equiv C \!-\! H$$

**42**

The intensities of ED peaks depend on the amplitudes of the molecular vibrations. In favorable cases this enables a study of the dynamics of intramolecular motion, allowing one, in turn, to determine force field parameters used in molecular mechanics, as discussed in Section 3.3. This application is of significant importance in view of the constantly growing applications of molecular modeling in chemistry discussed in Chapter 11.

To summarize, ED provides geometrical parameters and (together with the microwave spectra presented in Section 2.3.1) information on the molecular force field. Due to the proportionality of the scattered peak intensity to the molecular concentration, in favorable cases it is possible to determine the conformational equilibrium in the mixture. The temperature dependence of the equilibrium constant of these conformers could provide information on the thermodynamic functions of the molecule under study.

As discussed, the possibility of studying isolated molecules in the gas phase is the main advantage of ED over X-ray and ND. It is especially important in the case of molecules with low rotational barriers, where crystalline forces in the solid state can induce conformations different from those in the isolated molecule. The best-known example of this kind, biphenyl **35** [12], has been mentioned in Section 2.2.1. Another interesting example of the discrepancies between ED results on one hand and X-ray and ND on the other hand is provided by true ionic salts, for which interionic distances in the solid state and in the gas phase are different [28]. Yet another example of an advantageous application of ED, which falls outside the scope of the present monograph, is a study of free radicals, which do not exist in the solid state. ED represents one of the few possibilities of examining their spatial structure.

In some cases, bond lengths and angles determined by means of ED are more accurate than those obtained using X-ray techniques, since electrons are scattered by atomic nuclei, while X rays react to electron distribution. However, the application of ED is restricted to rather small molecules, and not all geometrical parameters can be obtained by its use, since it yields information on interatomic distances, not on the angles. The possibility of studying molecules in the gas phase ensures the complementarity of ED and microwave method. The latter one, briefly presented in Section 2.3.1, cannot be applied to nonpolar species that can be studied by ED. The accuracy and possibilities of ED and MW will be compared in Section 2.3.2.

## 2.3 Spectroscopic Methods

In these methods a substance under investigation in the gas phase, as a solution, or in the solid state interacts with electromagnetic radiation having the frequency characteristic of a given spectral technique. Most frequently, the absorption of radiation is studied. The division of electromagnetic radiation into regions was presented in Fig. 2.1. Electronic transitions typically correspond to an energy difference of ca. 500 kJ/mol, vibrational levels are separated by ca. 20 kJ/mol, while the energy difference between rotational levels is only ca. 0.05 kJ/mol (see Fig. 2.7 for a schematic representation of the levels). Therefore, the Born–

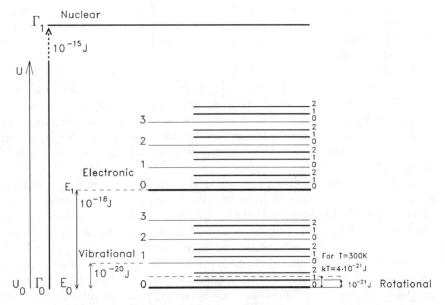

**Figure 2.7**   Schematic representation of molecular energy levels.

Oppenheimer approximation (discussed in Section 3.2.2 in some detail) separating electronic, vibrational, and rotational transitions was proposed to enable the solution of the Schrödinger equation. The limitations of the approximation will be shown below, but it should be stressed that the procedure cannot be used in the case of nonrigid molecules **21** and **23,** presented in the Introduction. It is also inapplicable in the narrow spectral range between 1 mm and 1 cm, since in this region all molecules behave as highly mobile nonrigid objects [29].

## 2.3.1 Microwave Spectra

The study of the absorption of radiation in the microwave region [30] is the main spectroscopic method of structure elucidation. It also provides the most accurate data on dipole moments and enables the analysis of internal rotations and inversional transitions. The study of microwave spectra in astronomy allows one to study organic molecules under conditions dramatically different from those in a chemical laboratory on Earth. This can lead to the discovery of new molecules. For example, cyanopolyynes **43** were first identified in MW spectra by astronomers on the basis of theoretical calculations [31]. The subsequent synthesis of these previously unknown substances proved their existence in the cosmos. Microwave spectra (MW) are essentially pure rotational absorption spectra in which the probability of a transition between different rotational levels is proportional to the square of the matrix element of the dipole moment. Hence the first limitation of the method: It can only be applied to molecules possessing a permanent dipole moment.

$$\text{H} - \left(\text{C} \equiv \text{C}\right)_n - \text{C} \equiv \text{N}$$

**43**

Molecular rotations can be most easily analyzed in the coordinate system associated with its axes of inertia. In such a system the rotational Hamiltonian (see Section 3.2.2 for a discussion of the separation of molecular rotations from vibrations and electronic transitions) can be expressed in the following form:

$$H_r = P_a^2/2I_a + P_b^2/2I_b + P_c^2/2I_c \tag{2.7}$$

where $I_a$, $I_b$, and $I_c$ are the moments of inertia and the *momentum operators P* are expressed in $h/2\pi$ units. The moments of inertia with respect to the main axes are equal to

$$I_a = \sum_i m_i\, r_{ia}^2 = \sum_i m_i(b_i^2 + c_i^2) \tag{2.8}$$

$$I_b = \sum_i m_i\, r_{ib}^2 = \sum_i m_i(a_i^2 + c_i^2) \tag{2.9}$$

$$I_c = \sum_i m_i\, r_{ic}^2 = \sum_i m_i(a_i^2 + b_i^2) \tag{2.10}$$

where $a_i$, $b_i$, and $c_i$ are the respective distances of the $i$th atom from the corresponding axis.

The rotational eigenproblem described by Eq. (2.7) is first solved in the rigid rotor approximation for certain specific cases such as diatomic molecules, linear molecules, and symmetric tops. Then the influence of vibrations on the determined moments of inertia is estimated. The moments are simple functions of bond lengths and angles in the molecule under investigation. For instance, for a diatomic or linear molecule $I_a = 0$, $I_b = I_c$, and the rotational axes are perpendicular to the molecular axis. In this case the energy associated with rotation $E_r = hBJ(J + 1)$, where $J$ is the rotational quantum number assuming the values 0, 1, 2, etc., and $B = h^2/8\pi I_b^2$. Due to the selection rule $\Delta J = \pm 1$, the rotational spectrum is very simple in this case; namely, it consists of a series of equidistant lines corresponding to the transition frequencies $\nu = 2B(J + 1)$. Knowing the distance between the lines, one can calculate the value of the $B$ constant, the moment of inertia $I_b$, and the length of the bond in the diatomic molecule under investigation.

In polyatomic molecules the determination of moments of inertia is more complicated. Moreover, the calculated moments are insufficient for the specification of all bond lengths and angles. As a rule, an additional assumption is made, namely, that an isotopic substitution changes only the mass of a molecule but not its geometrical parameters. On this basis so-called substitution structures $r_s$ are calculated.

As mentioned, MW is the only techniques enabling measurements of very small dipole moments inaccessible by dielectric techniques. For instance, theoretical calculations yielded a nonzero value of the ground-state electric dipole moment of methane due to centrifugal distortion during molecular rotations. The prediction was of the order of $10^{-5}$–$10^{-6}$ D. Such a small value could only be measured by MW, which yielded $(5.38 \pm 0.10) \times 10^{-6}$ D for the dipole moment of methane [32]. Thus, due to the very small distortions, even in the ground electronic and vibrational state methane does not have exact $T_d$ symmetry (see Chapter 4 for a discussion and designation of molecular symmetry). Of course, the effect is usually negligible, but is of importance in studies of intramolecular motion and of otherwise forbidden spectral transitions.

The potentialities and limitations of structural studies by means of microwave spectroscopy are as follows:

1.  As mentioned earlier, only molecules possessing a permanent dipole moment can be studied by this technique, but for them the method enables the most accurate studies of the moments.
2.  The method can be applied only to molecules having fewer than 20–30 atoms. As many as possible isotopically substituted molecules have to be studied simultaneously.
3.  The parameters of atoms lying far from rotational axes are estimated more precisely than the corresponding values for atoms lying close to the axes.
4.  MW spectra usually provide reliable information on the planarity of molecules under study, since for the planar ones a relation $I_a + I_b = I_c$ holds. Therefore, a nearly zero value of the inertia defect $\Delta = I_c - (I_a + I_b)$ proves the planarity.

5. Quadrupole splitting for nuclei with spin greater than $\frac{1}{2}$ exhibiting a nonuniform spread of nuclear charge complicates rotational MW spectra.
6. More complicated spectra are also observed for molecules in which internal rotation or inversion take place. The splitting due to these transitions can provide information on the corresponding barriers. The classical study of inversion in ammonia going through a planar transition state should be mentioned here [33].
7. The assumption of the independence of geometrical parameters on isotopic substitution is approximate only. It is valid for heavy atoms, but when a hydrogen atom is substituted for deuterium, the difference $r_z(CH) - r_z(CD) = 3 \times 10^{-13}$ m is comparable with the order of the accuracy of modern experimental bond length determinations. Such a difference can lead to inconsistencies in experimental results.

The division into pure rotational and vibrational spectra is not precise. Because of this, transitions between rotational levels can be also observed in rotation–vibrational spectra in IR and Raman spectroscopy. This enables one to study rotational spectra of highly symmetrical molecules having no dipole moment and to determine geometrical parameters of a molecule in excited vibrational states.

As mentioned, different physical phenomena are involved in different methods used for the determination of molecular structure, and the values measured by these methods refer to slightly different quantities. The most important parameters describing bond lengths determined by ED and MW methods are [28]:

$r_e$—equilibrium distance between positions of atomic nuclei corresponding to the potential function minimum. This value has to be compared with the results of quantum calculations.

$r_g$—average bond length, or, more generally, the average distance between nuclei at temperature $T$.

$r_\alpha^0$ and $r_z$—distance between average positions of nuclei in the ground vibrational state calculated from spectroscopic data (different from $r_g$).

$r_\alpha$—distance between average nuclear positions in thermal equilibrium at temperature $T$.

$r_a$—effective internuclear distance in the expression of the molecular contribution to electron scattering intensities.

$r_0$—effective internuclear distance determined from the rotational constants, usually referred to the ground vibrational state.

$r_s$—effective internuclear distance obtained from the isotopic substitution coordinates of the respective atoms.

$r_{av}$—average internuclear distance corrected for the effects of intramolecular motion. Usually obtained from a combined analysis of rotational spectra and ED data.

## 2.3.2 Comparison of the Results Obtained by ED and MW

Molecules in the gaseous state are studied by both methods. Therefore, a comparison is of interest. As mentioned, the accuracy of today's determination of geometrical parameters by means of various methods is so high that the differences in their results have become significant. For example, for acetone $(CH_3)_2CO$ [34] $r_g(C{=}O)$ is equal to $1.209(4) \times 10^{-10}$ m (ED), while the corresponding MW value $r_0$ is equal to $1.222(3) \times 10^{-10}$ m. A smaller but significant difference in values was determined by these methods for the length of the C–C bond $r_g = 1.515(3)$ and $r_0 = 1.507(3) \times 10^{-10}$ m. Such differences are not important for most synthetic chemists, but they gain in significance if one wants to study changes in bond lengths in a series of molecules induced by a substitution or ring size change. In such cases the results must be obtained by the same method, or the data should be recalculated into quantities independent of the methods used. This is especially important when the experimental data are compared with the results of quantum *ab initio* calculations, which significantly depend on the assumed geometry of a molecule under study. A very elegant example of the simultaneous use of various methods for the determination of the bond lengths in cubane will be discussed in Section 4.2.7.3 [35].

### 2.3.3 Infrared Spectra [36]

As discussed in detail in Section 3.2.2, in the Born–Oppenheimer approximation rotational and vibrational molecular motion can be separated to a good approximation. One can also treat vibrations within classical mechanics and add quantization afterwards. For the sake of simplicity, let us consider a vibrating diatomic molecule as a harmonic oscillator. It consists of atomic cores (that is, of atoms stripped of their valence electrons) with masses $m_1$ and $m_2$ situated at an equilibrium distance $r_e$ and connected by a spring with a force constant $f$. According to Hooke's law, a deformation of this system from the distance $r_e$ to the distance $r$ evokes a force acting on the masses $m_1$ and $m_2$ resisting the deformation from the equilibrium state. This force is proportional to the value $q = r - r_e$ and to the force constant $f$, but its sign is opposite to that of the deformation:

$$F = -fq \tag{2.11}$$

The minus sign describes the fact that the extended spring tends to shorten, while a compressed one tends to extend. The force $F$ can be expressed as a function of the potential energy of the spring:

$$F = -\frac{dV}{dq} \tag{2.12}$$

Then, the force constant

$$f = \frac{d^2V}{dq^2} \tag{2.13}$$

and kinetic energy of this molecule

$$T = \tfrac{1}{2}\,\mu\vartheta^2 \tag{2.14}$$

where the reduced mass $\mu$ by definition satisfies the equation $1/\mu = 1/m_1 + 1/m_2$ and the velocity $\vartheta$ is equal to the derivative $dq/dt$. The classical equation of motion of a harmonic oscillator can be shown to be

$$\mu\frac{d^2q}{dt^2} + fq = 0 \tag{2.15}$$

A solution of this equation describes the time dependence of the displacement $q$

$$q = Q \cos 2\pi\nu t \tag{2.16}$$

where the frequency $\nu$ is equal to $(1/2\pi)(f/\mu)^{1/2}$. In the quantum model the energy of an oscillator is described by Schrödinger equation

$$-\frac{(h/2\pi)^2}{2\mu}\frac{d^2\psi}{dx^2} + Vx = E\psi \tag{2.17}$$

Its solutions yield the energies of quantized vibrational levels:

$$E_{vib} = h\nu(\theta + \tfrac{1}{2}) \tag{2.18}$$

where $\nu$ is the frequency of the vibrations and $\theta$, which equals 0, 1, 2, etc., is vibrational quantum number. It is very important that, according to quantum calculations, *in the ground vibrational level for $\theta = 0$ the vibrational energy is not equal zero* but $\tfrac{1}{2}h\nu$. In the harmonic approximation the distances between consecutive levels of the oscillator are equal, and a selection rule allows only for transitions between neighboring levels. In an anharmonic approximation the distances diminish with increasing vibrational quantum number, and the selection rule does not hold strictly. In this case $\Delta\theta$ may be equal to $\pm2$, $\pm3$, etc., but the intensities of these overtones (corresponding to multiplicities of the fundamental frequencies) are much lower.

In polyatomic molecules vibrations are usually not localized in a bond, and their only precise classification is based on the molecular symmetry discussed in Chapter 4. However, in many cases one can, to a good approximation, analyze vibrations localized in some molecular fragments. Thus one speaks about stretching and bending of bonds and about torsions around bonds. Such an approximation holds best for the vibrations involving light hydrogen atoms oscillating fairly independently of the rest of the molecule under study. Thus one speaks of the stretching vibration of a CH bond, $\nu_{CH}$; of the bending of a CCH fragment, $\vartheta_{CCH}$; of the torsional vibrations $\delta_{CH}$ in an aromatic ring, etc. These are so-called characteristic vibrations, allowing one to identify these fragments in a molecule. In any case it should be stressed that the only precise classification of molecular vibrations is that based on their symmetry. The assignment of molecular vibrations is very important for the elucidation of molecular structure [37, 38]. It is a complicated task that will not be discussed here.

For polyatomic molecules, instead of Eq. (2.15), vibrations are described by a matrix equation. In this equation the reduced mass $\mu$ is replaced by kinetic energy matrix consisting of elements that are complicated functions of the relevant atomic masses and bond lengths and angles. Similarly, the force constant $f$ in this equation is replaced by force constant matrix **F**. Knowing the constants and molecular geometry, one can calculate its normal vibrations. Such a procedure is helpful when a molecule under investigation exists as a mixture of different conformers having the same force field but different spatial structures. This is the case for chlorocyclohexane **24** [39], where there are two conformers with equatorial and axial orientations of the carbon–halogen bond, yielding two $\nu_{CCl}$ bands in the infrared region. The frequencies of these bands can be calculated using a known force field of this molecule. On the other hand, a solution of the so-called inverse problem, that is, establishing force constants on the basis of known spectra and geometrical parameters, is a much more difficult task. From the mathematical point of view, the problem is ill-defined, since, with the exception of very small highly symmetrical molecules, the number of force constants that are sought is larger than the number of normal vibrations yielding the equations from which the constants can be determined. Additional equations can be used when the spectra of isotopically substituted molecules are at our disposal (see Section 2.3.1) (or when additional information on Coriolis interactions and centrifugal distortions is provided [40]), but despite their inclusion the number of unknown force constants is still too large. Thus transferability of the force constants of molecular fragments in different molecules is assumed, and as a first approximation one takes the constant values from similar molecules. Then the frequencies are calculated and the constants changed to obtain the best reproduction of the experimental spectrum. This process is repeated until an assumed difference between the calculated and measured spectrum is obtained. Such an iterative procedure is called the *self-consistent field* (SCF) method. It is used not only in IR but also in quantum mechanics and molecular mechanics, to be discussed in Chapter 3. The assumption on the transferability of force fields for fragments is equivalent to the more general hypothesis of the additivity of molecular fragments that is frequently used in organic chemistry. In agreement with this hypothesis, many quantities such as the dipole moment, enthalpy of formation, or steric energy can be estimated as a sum of terms describing the respective contributions of the constituent bonds or atoms.

Absorption of IR radiation can be shown to be proportional to the square of the change of the dipole moment during a vibration $d\mu/dq$. Therefore, vibrations during which the dipole moment does not change are infrared inactive. They correspond to so-called *forbidden transitions*. The latter can be determined on the basis of group theory [41] by analyzing the symmetry of the vibrations and that of the transition moment. The following rationale shows the reasoning without the involvement of tedious mathematics. For the linear molecule $O_2$ the dipole moment change during the vibration is equal to zero; thus the molecule does not exhibit IR spectra. Similarly, other molecules possessing an inversion center do not have symmetric IR absorption bands. Together with information from the corresponding Raman spectra, which will be presented in the next section, it enables one to determine the symmetry of such molecules. This and the study of different conformers are the main source of information on the spatial structure of molecules provided by IR spectra. Let us look at few examples of such applications:

1. Vinyl ketones can exist as a mixture of *s-trans* and *s-cis* conformers **44a** and **b** ($R$ = Me, $X$ = H), differing in the mutual orientation of the double bonds. The isomerism manifests itself in the splitting of a band at $1680-1705$ cm$^{-1}$ usually but not precisely ascribed to the stretching vibrations of the carbonyl group $\nu_{C=O}$. The relative intensity of the components of this band depends on many

factors, but they were empirically found to correlate with the volume of the substituent $R$ on the carbonyl group. For instance, the intensities corresponding to the stretching vibrations of *s-cis* to *s-trans* conformers were equal to 154:227 for $R$ = Et, 263:189 for $R$ = *iso*-propyl, and 403:44 for $R$ = *tert*-butyl [42]. This finding is in agreement with model considerations yielding a larger amount of the *s-cis* conformer for voluminous substituents due to nonbonding repulsions.

2. Only one of two diols **45** and **46** under investigation can form an intramolecular hydrogen bond [43]. Thus they can be differentiated by studying IR spectra in the region corresponding to the stretching vibrations of the OH group.

3. The nonplanar structure of the cyclobutane ring **47** was established on the basis of a splitting of the band corresponding to the $\nu_{CD}$ stretching vibration in the spectrum of the monodeuterated compound [44]. Otherwise, only one band would be present in the 2000–2500 cm$^{-1}$ region.

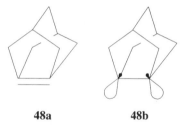

| trans-s-trans | trans-s-cis | cis-s-cis | cis-s-trans |
|---|---|---|---|
| **44a** | **44b** | **44c** | **44d** |

**45**          **46**          **47**

An interesting possibility is provided by measurements of low-temperature IR spectra in noble gas matrices, enabling the observation of short-lived species [45]. The next two examples make use of this possibility.

4. The nonplanar distortion of the double bond in the tricyclic compound **48a** is so significant that the formula **48b** was invoked to describe its structure. In view of the instability of **48,** only IR spectra at low temperature in a matrix could be

**48a**          **48b**

studied [46]. The typical frequency $\nu_{C=C}$ of the stretching vibration is ca. 1650 cm$^{-1}$, while the corresponding value for **48** decreased to the highly unusual value of 1496 cm$^{-1}$ due to the strongly diminished C=C double bond character resulting from its nonplanar distortion.

5. As discussed in detail in Section 3.2.5, the structure of an unstable $C_6S_6$ molecule **49** could be studied only by matrix isolation IR and by theoretical calculations. The latter were carried out for five structures **49a–49e**. A simultaneous use of both methods allowed one to propose **49d,f** as the most stable structures [47].

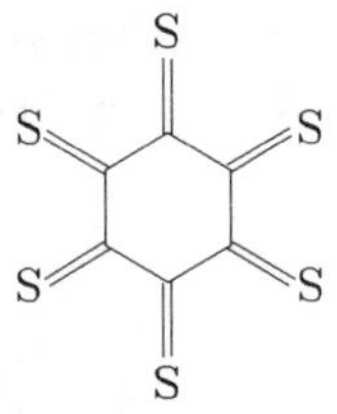

49a 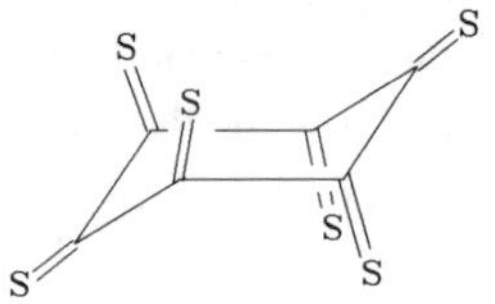          49b 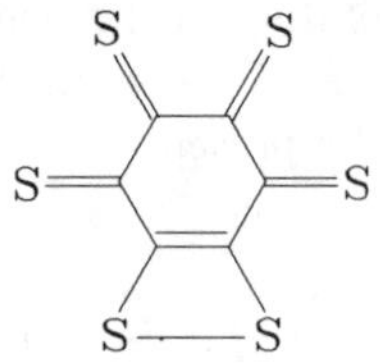          49c

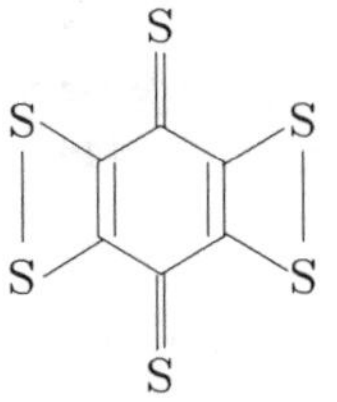

49d 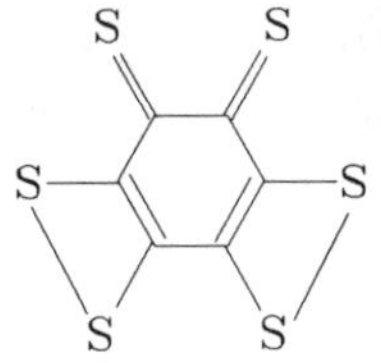          49e 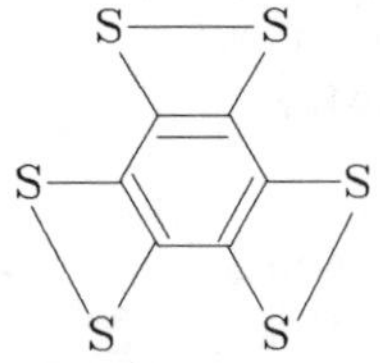          49f

## 2.3.4 Raman Spectra

Instead of absorption of the corresponding IR radiation, scattering of visible or ultraviolet radiation by a vibrating molecule is observed in the Raman spectrum [36]. In the Raman case the frequency difference between the incident and scattered waves is equal to the frequency of a normal vibration of the molecule $\nu$. For this reason frequencies in IR and Raman spectra are very close, but their intensities differ. As stated in the former section, the IR intensity is proportional to the change in dipole moment during a vibration, while that in Raman spectra is proportional to the square of the change in polarizability $d\alpha/dq$ during the vibration. The polarizability $\alpha$ is a tensor quantity characterizing the mobility of electrons in an electric field. It assumes the form

$$\alpha_{\mu\nu} = \begin{vmatrix} \alpha_{xx} & \alpha_{xy} & \alpha_{xz} \\ \alpha_{yx} & \alpha_{yy} & \alpha_{yz} \\ \alpha_{zx} & \alpha_{zy} & \alpha_{zz} \end{vmatrix}$$

and is subjected to selection rules different from those governing IR spectra. For example, as mentioned in the former section, totally symmetric vibrations of molecules possessing an inversion center that are inactive in IR spectra are active in the Raman case. On the other hand, nonsymmetric vibrations for such molecules are IR active but Raman inactive. This constitutes the so-called *mutual exclusion principle,* allowing one easily to establish the existence of an inversion center in a molecule under investigation. The selection rules for IR and Raman spectra for benzene, a molecule of $D_{6h}$ symmetry (see Chapter 4 for the symmetry notation), can serve as an example of such complementarity of these spectra [48].

### 2.3.5 A Comparison of IR and Raman Spectroscopy

As noted previously, the frequencies measured in IR and Raman spectra are very close. Therefore, the former method is mostly used for standard identification purposes. The latter one is applied in special cases, such as studies of biomolecules in water solutions, which are nontransparent to infrared radiation, in studies of weak infrared bands such as those corresponding to stretching vibrations of the C≡C bond, and in studies involving the mutual exclusion principle.

### 2.3.6 Ultraviolet and Visible Spectra [49]

Spectra in the ultraviolet–visible (UV–VIS) region are due to electronic transitions. From the point of view of their applications in organic chemistry, the most interesting is the study of conjugated systems of double bonds and of aromatic systems. A molecular fragment that to a good approximation can be considered responsible for absorption in these regions is called a *chromophore.* Ethylene and carbonyl groups as well as aromatic rings are the most common examples of chromophores. UV–VIS spectroscopy is a mainly analytical technique since (1) the spectra usually do not exhibit a pronounced dependence on molecular spatial structure, and (2) a full interpretation of electronic transitions is impossible without highly accurate quantum calculations.

Without the application of tedious calculations, UV–VIS spectra enable a determination of the existence of conjugated bond systems. Sometimes they provide information on the position of substitution or protonation. In supramolecular chemistry (Chapter 10) they are applied to determine the stoichiometry of the complexes under investigation. However, the use of these spectra to determine the spatial molecular structure is very limited. UV–VIS spectra correspond to transitions between the ground and excited electronic states. As such, they depend simultaneously on both the ground- and excited-state geometries, and the elucidation of the spatial structure in the ground state from UV–VIS spectra is not straightforward. It seems that the application of the spectra to these purposes will gain in importance in the future. The recent synthesis of molecules having highly distorted double bond systems such as [6]-layered cyclophane **10**, $C_{60}$ **11**, and permethylated benzene **38,** as well as the development of computers enabling accurate calculations of electronic transitions of fairly large molecules, will foster the application of this technique with this purpose in mind.

## 2.3.7 Circular Dichroism and Optical Rotatory Dispersion [50–52]

Both circular dichroism (CD) and optical rotatory dispersion (ORD) techniques make use of the interaction of polarized light with a chiral molecule. Let us remember that molecular chirality was introduced in Chapter 1. Its detailed discussion with all pertinent definitions will be presented in Chapter 5. Polarized light interacts in a different way with the enantiomers, that is, mirror images, of a chiral molecule, enabling the determination of absolute configuration with help of empirical correlations.

### 2.3.7.1 Polarization of Light

The propagation of linearly polarized light in the direction of the $z$ axis is associated with cosine changes in electric $\mathbf{E}$ and magnetic $\mathbf{H}$ vectors perpendicular to the direction of the propagation. The electric vector exhibits a cosine dependence on time, while the magnetic vector dependence is as a sine. Therefore, the electric vector

$$E = E_0 \cos(2\pi\nu t - 2\pi z/\lambda) = E_0 \cos \omega(t - z/c_0) \qquad (2.19)$$

where $\nu$ is the frequency of radiation, $\lambda = c_0/\nu$ denotes wavelength, and $\omega = 2\pi\nu$. *Natural light* is composed of the radiation of various wavelengths. It does not have one plane of polarization; that is, the light does not display a single plane of oscillation of the electrical vector $\mathbf{E}$. With the use of monochromators and polarizers, one can obtain linearly polarized light, for which the electric vector $\mathbf{E}$ oscillates in a plane. Simultaneously, the magnetic vector $\mathbf{H}$ executes oscillations in the perpendicular plane. Thus linearly polarized light can be treated as a superposition of two waves circularly polarized in opposite directions. In such light the electric vector does not oscillate in one plane, but its end evolves, forming a screw with a constant pitch on the surface of a cylinder. The magnetic vector is always perpendicular to the electric, executing the same evolution. A sum of two waves of the same frequency and different intensities circularly polarized in the opposite directions yields elliptically polarized light.

### 2.3.7.2 The Interaction of Light with an Achiral Molecule

The oscillating electric vector of incident light causes vibrations of the molecular electron shell, producing in this molecule an induced dipole moment, while opposite vibrations of much heavier nuclei can be neglected. In classical mechanics such a vibrating dipole is a source of electromagnetic radiation. By an elastic collision of a photon with the molecule, the whole energy obtained is emitted, while by inelastic scattering a part of the energy of radiation is absorbed by the molecule and changed to thermal energy. The equation

$$\Delta E = h\nu \tag{2.20}$$

where $\Delta E$ is the energy difference between two electronic levels, is, of course, the necessary condition for radiation with frequency $\nu$ to be absorbed. However, it is not the sufficient condition. To realize the transition, its electric moment $\mu$ has to be different from zero.

Because of internal vibrations and rotations as well as collisions with the solvent molecules, solution spectra exhibit broad bands instead of narrow lines. The intensity of these bands, equal to the surface under absorption curve, is proportional to the square of the transition moment $\mu$

$$\int\epsilon \, d\lambda \propto \mu^2 \tag{2.21}$$

where $\epsilon$ is the absorption coefficient and $\lambda$ is the wavelength.

### 2.3.7.3 The Interaction of Light with Chiral Molecules

The interaction between right and left hands differs from that between two right hands. Similarly, the interaction of two circularly polarized components of linearly polarized light with a chiral molecule is different. If a substance under investigation consists exclusively (or mainly) of molecules with one absolute configuration, then in its absorption region there are two different values of the refraction index for these components $n_L$ and $n_R$ and different velocities of propagation of the waves circularly polarized in opposite directions. Due to the differences in velocities, the waves that have passed a layer containing molecules of a definite configuration (or at least an excess of them) sum to the linearly polarized light, with the plane of polarization twisted with respect to the plane of incident light. The twist angle depends on the thickness of the layer $l$, the difference of refraction indices $\Delta n = (n_L - n_R)$, and the wavelength $\lambda_0$

$$\alpha^0 = 1800l \, \Delta n/\lambda_0 \tag{2.22}$$

In turn, $\Delta n$ depends on the wavelength. The value $\Phi = \alpha/cl$, where $c$ denotes concentration and $l$ the layer thickness is called *optical rotatory dispersion*. If, in addition to $n_L \neq n_R$, the absorption coefficient $\epsilon_L \neq \epsilon_R$, then after passing through an absorbing layer containing chiral molecules of one absolute configuration (or an excess thereof), the linearly polarized waves sum to an elliptically polarized one. The wavelength dependence of $\Delta\epsilon = \epsilon_L - \epsilon_R$ describes *circular dichroism*. Both processes ORD and CD are jointly known as the *Cotton effect*, and the substances exhibiting this effect (as well as some others, such as polarization effects in Raman spectra) are called *optically active* compounds. Optical activity is characteristic for chiral molecules possessing inherently chiral chromophores or achiral chromophores with induced chirality (Fig. 2.8).

# Inherently chiral chromophores

# Inherently achiral chromophores

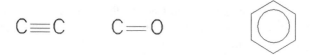

**Figure 2.8**    The classification of chiral chromophores.

### 2.3.7.4 Model Description of Optical Activity

Optical activity is associated with the simultaneous interaction of electromagnetic radiation with a chiral molecule. The magnitude of the interaction is related to rotational strength

$$R = \boldsymbol{\mu} \cdot \mathbf{m} \, \cos(\boldsymbol{\mu} \cdot \mathbf{m}) \tag{2.23}$$

where $\boldsymbol{\mu}$ is an electric vector of the transition and $\mathbf{m}$ is the corresponding magnetic vector induced by the changing electric field. The direction of the $\mathbf{m}$ vector can be determined from the *right thumb rule*, which states that electric current propagating in the direction of the forefinger of the right hand is accompanied by a magnetic field described by the vector $\mathbf{m}$ with a direction given by the thumb. Both electric and magnetic moments must change to observe optical activity.

A nonzero value of the magnetic transition moment accompanies light absorption if the electronic transition causes a rotation of electric charges corresponding to a circular current. To determine the magnitude of the moment $\mathbf{m}$, one considers the initial and final orbitals of the transition. Then a rotation of the former by a smaller angle has to be executed in such a way that it overlaps the final orbital, with the signs of the orbital lobes coinciding. Since circular dichroism takes place when rotational strength assumes a nonzero value, then not only must $\mu$ and $\mathbf{m}$ vectors differ from zero but the angle they form is of importance. If the vectors form an acute angle, then CD is positive; when the angle is obtuse, the CD is negative. No effect is observed when the vectors $\mu$ and $\mathbf{m}$ form a right angle even for nonzero values of the vectors.

Complicated quantum calculations are necessary to determine rotational strength, but for many practical applications the sign of the rotational strength is sufficient for a configurational assignment to be made. In many cases the sign can be deduced on the basis of simple reasoning from the direction of the $\mu$ and $\mathbf{m}$ vectors. Let us consider a disulfide fragment of 3-dithia-$\alpha$-steroid **50,** an example taken from Ref. 51. $n^+$ and $n^-$ combinations can be formed from nonbonding orbitals $n_s$ on sulfur atoms (see Chapter 3 for a discussion of energy levels obtained in quantum calculations). The latter higher-energy orbital represents the highest occupied molecular orbital (HOMO). The lowest unoccupied MO (LUMO) is a $\sigma^*$ orbital, also presented in Fig. 2.9. In the steroid under consideration CS bonds form a positive angle around the SS bond. The charge distributions in the HOMO and LUMO as well as the corresponding $\mu$ and $\mathbf{m}$ vectors shown in the figure indicate that the given conformation around the SS bond is responsible for positive CD. It is very important that the effect is associated with the absolute conformation of the CSSC fragment. To a first approximation, it does not depend on the configuration of the rest of the molecule, which has to be established by different methods. It should be stressed that asymmetric atoms in the vicinity of a chromophore (C5, C10, etc.) exert no direct influence on the CD sign. The influence is indirect, since the configuration at these atoms controls the mutual orientation of the CS bonds through nonbonded interactions.

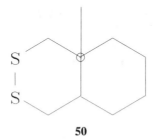

**50**

A mutual orientation of $\mu$ and $\mathbf{m}$ vectors in the case of achiral chromophores in a chiral environment can be found in an analogous way. In many cases it is impractical to analyze such transitions in a qualitative way, and complicated quantum calculations are often demanding. Therefore, empirical rules are often used, such as the *octant rule* for carbonyl groups, allowing one to determine in what part of the

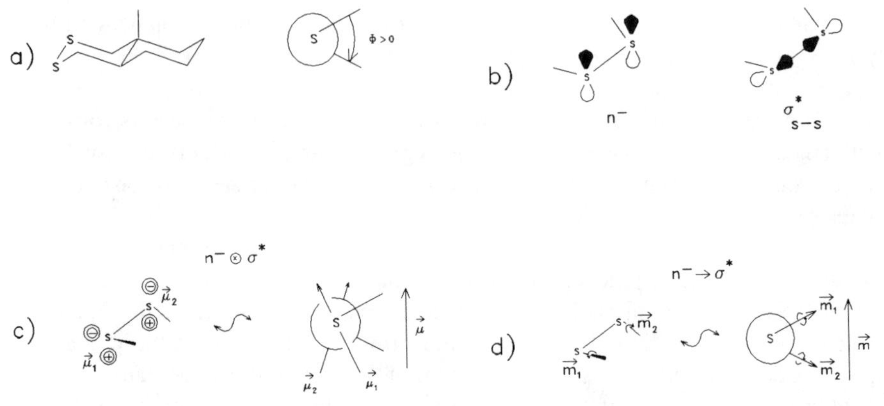

a)   b)   

c)   d)

$$\vec{\mu} \uparrow \vec{m} \Rightarrow \Delta\varepsilon > 0$$

**Figure 2.9** The mutual orientation of (a) CS bonds, (b) $n^-$ and $\sigma^*$ orbitals on S atoms, (c) electric $\mu$ and (d) magnetic $m$ dipole moments in **50**.

space around this group the substituents are situated [52]. A few examples of the application of CD spectra for an elucidation of spatial structure are:

1. In addition to $E$–$Z$ isomerism, substituted N-nitrosopyrrolidines **51** are known to assume nonplanar, flexible conformations of five-membered rings. Due to the existence of these latter conformations, CD spectra of these compounds exhibit a considerable temperature dependence, as shown in Fig. 2.10 [53].

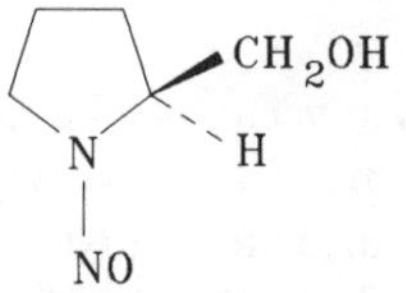

**51**

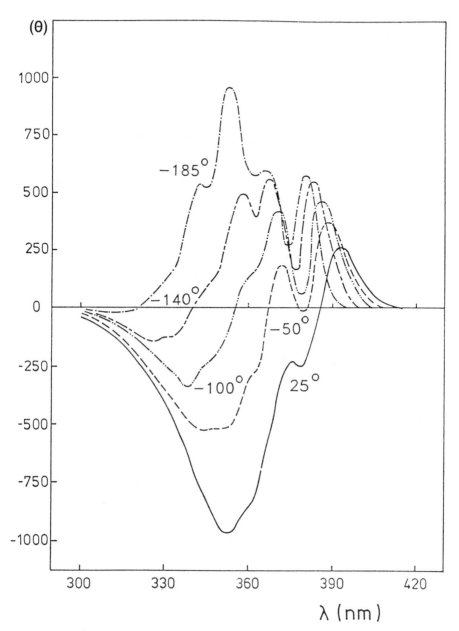

**Figure 2.10** The temperature dependence of CD spectra of N-nitrosopyrrolidine **51.** (Courtesy of Dr. T. Polonski.)

2. Polypeptides are known to exist in the form of ordered ($\alpha$-helix and $\beta$-sheets) spatial arrangements or as random coils, determining their secondary structure. Greenfield and Fasman have shown for poly-$\alpha$-lysine that these arrangements, induced by solvent of pH changes, influence the CD spectra [54].

3. As said before, an achiral chromophore such as carbonyl group can exhibit chirality in a chiral environment. For this reason CD and ORD techniques are the most widely used methods in studies of configurational assignments of organic molecules. Chiral perturbation can be situated outside the achiral molecule under investigation. This is often the case in supramolecular chemistry where Cotton effect is due to the effect of chiral host molecules on achiral guest ones (see Chapter 10 for the definition of host and guest molecules). For instance, the complex of benzo[a]pyrene **52** with $\gamma$-cyclodextrin **14c** has a small but measurable *induced chirality* in spite of the planar structure of **52** apparently excluding chirality [55].

**52**

## 2.3.8 Nuclear Magnetic Resonance

NMR spectroscopy [56–59] is now one of the most used structure determination methods. It provides the most accurate information on molecular geometry in solution, especially on proteins and supramolecular systems. Therefore, in spite of the limited space in this book, NMR will be presented in some detail. Its foundations will be briefly sketched first, and the kind of information it yields will be discussed next. Numerous applications will accompany the presentation of this rapidly developing method.

### 2.3.8.1 Foundations of the Method

Many nuclei possess a spin nuclear momentum characterized by the spin quantum number $I$, which can assume only the values equal to a multiplicity of $\frac{1}{2}$. As discussed in the next chapter, there is no classical analogy to this property. $I$ is equal to zero for $^{12}C$ and $^{16}O$, and these nuclei do not exhibit NMR absorption. (As will be shown, this considerably facilitates the interpretation of the proton and carbon spectra of organic molecules.) Protons, $^{1}H$, as well as $^{13}C$, $^{15}N$, $^{19}F$, and $^{31}P$ nuclei have nuclear spin $I = \frac{1}{2}$, for deuterium, $^{2}H$, and $^{14}N$ the spin is equal to 1, etc. Spin

**Table 2.2** Magnetic Properties of the Most Important Nuclei

| Nucleus | Natural abundance (%) | Spin moment ($I$) | Magnetic moment ($\mu_N$) | Quadrupole moment ($e\ 10^{-28}\ m^2$) | Magnetogyric ratio ($10^7 T^{-1}\ s^{-1}$) |
|---|---|---|---|---|---|
| Neutron | — | $\frac{1}{2}$ | $-1.91304$ | 0 | $-18.325$ |
| $^1$H | 99.9844 | $\frac{1}{2}$ | 2.79285 | 0 | 26.725 |
| $^2$H | $1.56 \times 10^{-2}$ | 1 | 0.85738 | $2.77 \times 10^{-3}$ | 4.106 |
| $^7$Li | 92.57 | $\frac{3}{2}$ | 3.2560 | $-4.2 \times 10^{-2}$ | 10.396 |
| $^{11}$B | 81.17 | $\frac{3}{2}$ | 2.6880 | $3.55 \times 10^{-2}$ | 8.583 |
| $^{12}$C | 98.892 | 0 | 0 | 0 | 0 |
| $^{13}$C | 1.108 | $\frac{1}{2}$ | 0.70216 | 0 | 6.726 |
| $^{14}$N | 99.635 | 1 | 0.40557 | $2 \times 10^{-2}$ | 1.933 |
| $^{15}$N | 0.365 | $\frac{1}{2}$ | $-0.28304$ | 0 | $-2.711$ |
| $^{16}$O | 99.757 | 0 | 0 | 0 | 0 |
| $^{17}$O | $3.7 \times 10^{-2}$ | $\frac{5}{2}$ | $-1.8930$ | $-4 \times 10^{-3}$ | $-3.627$ |
| $^{19}$F | 100 | $\frac{1}{2}$ | 2.6273 | 0 | 25.166 |
| $^{23}$Na | 100 | $\frac{3}{2}$ | 2.2161 | 0.1 | 7.076 |
| $^{31}$P | 100 | $\frac{1}{2}$ | 1.1305 | 0 | 10.829 |
| $^{35}$Cl | 75.4 | $\frac{3}{2}$ | 0.82089 | $-7.97 \times 10^{-2}$ | 2.621 |
| $^{37}$Cl | 24.6 | $\frac{3}{2}$ | 0.68329 | $-6.21 \times 10^{-2}$ | 2.182 |
| $^{55}$Mn | 100 | $\frac{5}{2}$ | 3.4610 | 0.5 | 6.630 |
| $^{127}$I | 100 | $\frac{5}{2}$ | 2.7939 | $-0.75$ | 5.352 |

values and some other properties of the most important magnetic nuclei are collected in Table 2.2. The projection of the spin $I$ on an arbitrary axis can assume only the values

$$m_I = I, I - 1, I - 2, \ldots, -I$$

A nucleus with nonzero spin has a magnetic moment that can take only $2I + 1$ orientations with respect to a static magnetic field $B_0$. The energy of interaction of the induced magnetic moment of the spin $I$ with the field depends on their mutual orientation

$$E_{m_I} = -\mu_z B_0 = -\gamma(h/2\pi)B_0 m_I \tag{2.24}$$

where $\mu_z$ is the component of the magnetic moment on the $z$ axis parallel to the field $B_0$, and $\gamma$ is an empirical factor called the magnetogyric ratio of the nucleus. Thus in the magnetic field the nuclear energy levels are split. As schematically presented in Fig. 2.11, the splitting is proportional to the field $B_0$. The levels are populated according to the Boltzmann Law

$$N_1/N_2 = \exp(-\Delta E_{1,2}/kT) \tag{2.25}$$

where $N_i$ is the population of the level $i$, $\Delta E$ is the energy difference between the levels, and $k$ and $T$ are the usual Boltzmann constant and absolute temperature. It is important that the energy difference between nuclear levels is small. As a result, the population difference between the nuclear states is also small. The population excess at the lower state, determining the probability of a transition and, in turn, the

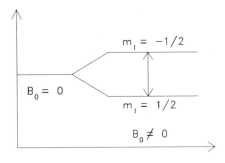

**Figure 2.11**   Schematic representation of the splitting of spin levels for $I = \frac{1}{2}$ in the magnetic field $B_0$.

sensitivity of the experiment, is of the order of $10^{-6}$. By increasing the magnetic field strength, the sensitivity of NMR increases.

A transition between the levels is induced by radio-frequency radiation $\nu$ propagating perpendicular to the direction of the static magnetic field. In a magnetic field, the magnetic moment of a nucleus with spin $I$ executes a precessional motion around the field direction, with the Larmor frequency $\nu$ proportional to the strength of this field $B_0$

$$\nu = \gamma B_0 / 2\pi \quad \text{or} \quad \omega = \gamma B_0 \tag{2.26}$$

The transitions between nuclear spin levels cause an equilibration of the populations of the lower and upper nuclear spin states. This leads to a saturation of absorption, and the system stops to absorb the incoming radiation. Due to thermal collisions of the nuclei (spin-lattice relaxation characterized by the $T_1$ time constant) and exchange of the spins between the neighboring atoms (spin–spin relaxation characterized by the $T_2$ constant), the nuclei return to the lower state, enabling further absorption. Spin-lattice relaxation is due to fluctuating magnetic fields, which induce transitions between nuclear levels. Out of several possible mechanisms of such fluctuating fields, dipolar relaxation, which makes use of a modulation of dipolar spin–spin coupling, provides valuable information. The study of spin-lattice relaxation times $T_1$ in $^{13}C$ spectra allows one to investigate molecular motion and intramolecular mobility. For instance, $T_1$ values for C1 and C4 carbon atoms in *para*-substituted phenols **37c** are larger than the corresponding values for *ortho*-**37a** and *meta*-positions **37b**, since the *para* molecule rotates preferentially around its longer axis [59]. The dipolar relaxation rate $R_1^{DD} = (1/T_1)^{DD}$ was shown to be proportional to $r^{-6}$ and the correlation time $\tau_c$. The latter quantity describes the reorientation of a molecule in a liquid, and $r$ denotes the distance between the nuclei. Therefore, measurements of dipolar relaxation rate can provide information not only on molecular motion but also on bond lengths.

The magnetic field within a molecule is smaller than the applied external field, since nuclear spins are *shielded by surrounding electrons*. Therefore, resonance conditions for nuclei in different environments are slightly different, and there is no absolute scale of *chemical shifts*. In spite of the fact that the shifts due to the differences in environments within a molecule are very small, they are characteristic of the environments, and their tabulated values are invaluable for analytical purposes. As with the other spectral techniques, we will not present such NMR applications here. In the following we will be interested in more subtle information on the spatial structure of molecules, which can be determined from NMR spectra. Some more sophisticated NMR techniques that allow one to assign the signals in the spectra will also be briefly presented. In the spectra chemical shifts of the substance under investigation $v_{subst}$ are measured with respect to the shift of a standard, reference compound, tetramethylsilane (TMS) $v_{ref}$ in case of $^1H$ and $^{13}C$ spectra. The shifts are usually denoted δ and expressed in parts per million (ppm):

$$\delta = (v_{subst} - v_{ref})/v_0 \tag{2.27}$$

where $v_0$ is the operating frequency of the spectrometer. Every independent group of protons (or other nuclei possessing nonzero spin) gives rise to a separate signal in the NMR spectrum. Thus the proton spectrum of ethyl benzene, $CH_3CH_2C_6H_5$, **53** presented in Fig. 2.12 reveals three groups of signals corresponding to methyl, methylene, and aromatic ring protons with relative intensities of 3:2:5, reflecting the number of protons of different kinds. (Contrary to the situation in proton spectra, relative intensities in standard $^{13}C$ spectra cannot be used to determine the number of carbon nuclei in different environments.) The following convention was adopted to describe the signals in an NMR spectrum. Nuclei in a molecule are denoted by capital letters, with subscripts giving their number. If their chemical shifts differ considerably, then the letters are taken from the beginning and the end of the alphabet. With intermediate differences (with chemical shift differences in Hertz comparable to spin–spin coupling constants, see the following) $K, L, M$ symbols are applied to nuclei. According to this convention, the ethanol spectrum is of the $A_3B_2X$ type.

$CH_2CH_3$

**53**

The methyl and methylene signals in the spectrum presented in Fig. 2.12 are split due to *spin–spin coupling*, since neighboring magnetic nuclei induce minor variations of the effective magnetic field acting on the nuclei. This is the so-called indirect or scalar spin–spin coupling, transmitted through covalent bonds and causing a splitting of the NMR signals into multiplets in solution spectra. Direct or dipolar coupling between nuclear spins through space is observed only in solid-state NMR. The *spin–spin coupling constant J* is a distance (in frequency units Hz) of two subsequent components of a multiplet. The $J$ value does not depend on the field $B_0$. In high-resolution NMR spectra one can observe the coupling constants between

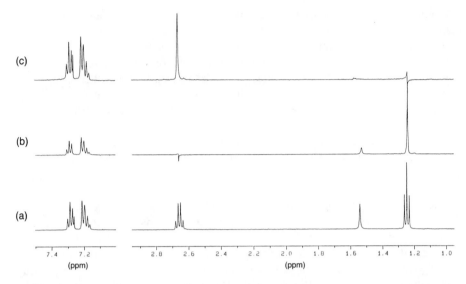

**Figure 2.12** (a) $^1$H NMR spectra of ethyl benzene. The signal at ca. 1.55 ppm is due to traces of water. (b) and (c) The same spectrum under decoupling of methylene and methyl groups, respectively. (Courtesy of Mr. J. Sitkowski.)

nuclei forming a bond, such as $J_{CH}$, those between nuclei separated by two or three bonds, $^2J$ or $^3J$, and, usually small, long-distance coupling through four or five bonds, $^4J$ or $^5J$, respectively. Such long-distance couplings are non-negligible only if the bonds are coplanar and form a so-called zig-zag arrangement.

The classification of spin–spin couplings is summarized in Table 2.3. The *multiplicity of a signal* is due to the spin–spin coupling. If there is no splitting, one speaks of a singlet (*s*). When a signal is split into two, three, four, six, or seven equidistant components, then we deal with a doublet (*d*), triplet (*t*), quadruplet (*q*), quintet (*qui*), sextet (*sxt*), or septet (*sep*), respectively. When there are different coupling constants in the system, then a more complicated picture (shown in Fig. 2.12 for the aromatic proton signals of **53** at ca. 7.2–7.3 ppm) arises.

If a coupling constant $J$ is small in comparison to the difference in chemical shifts, then the simplest first-order multiplets appear in the spectrum. In case of $m$ nuclei $A$ coupled with $n$ nuclei $X$ ($A_mX_n$ system) with spin equal to $\frac{1}{2}$, there is a splitting into ($n + 1$) components with intensity ratios given by the Pascal triangle presented in Fig. 2.13. In such a case the multiplet picture is especially simple. In the spectrum of ethyl benzene in Fig. 2.12(a), the spin moment of the methylene group (equal to $2 \times \frac{1}{2} = 1$) can orient itself with respect to magnetic field $B_0$ in a parallel, perpendicular, and antiparallel way, producing splitting of the methyl protons' signal at ca. 1.25 ppm into three components of 1:2:1 intensity. Analogously, the methyl protons ($\Sigma I = \frac{3}{2}$) cause the splitting of the methylene signal at ca. 2.65 ppm into four components with relative intensities of 1:3:3:1. The distance between the neighboring components of both multiplets yields the value of the spin–

**Table 2.3**  Classification of Spin–Spin Couplings

| Type of coupling | Number of separating bonds | Classification | Symbol |
|---|---|---|---|
| 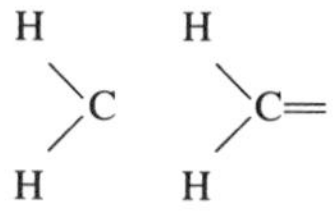 H     H<br>  \C  \C=<br>  /   /<br>H     H | 2 | Geminal | $^2J$ |
| 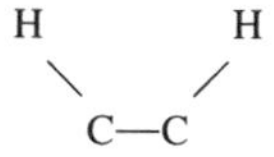 H      H<br> \   /<br>  C—C | 3 | Vicinal | $^3J$ |
| 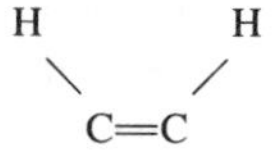 H      H<br> \   /<br>  C=C | 3 | Vicinal | $^3J_{cis}$ |
| 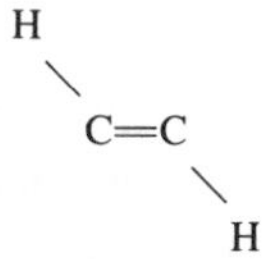 H<br> \<br>  C=C<br>     \<br>      H | 3 | Vicinal | $^3J_{trans}$ |
| Long-range couplings<br>H—C—C=C—H | 4 | Allylic | $^4J$ |
| H—C—C=C—C—H | 5 | Homoallylic | $^5J$ |

spin coupling constant $J$. In more complicated spectra the presence of equidistant splittings of different signals facilitates spectral assignments (see, for instance, Fig. 2.16).

Let us look at the difference between $AX$ and $AB$ spin systems presented schematically in Fig. 2.14. If the coupling constant is much smaller than the difference in chemical shifts, then two doublets with equal line intensities appear in the first spectrum. Otherwise, as can be seen from the second spectrum in Fig. 2.14, the intensities are perturbed, and the so-called "roof effect" manifests itself.

```
                     1
                  1  2  1
               1  3  3  1
            1  4  6  4  1
         1  5  10 10  5  1
      1  6  15 20 15  6  1
```

**Figure 2.13**  Pascal triangle representing intensities of the components of first-order multiplets.

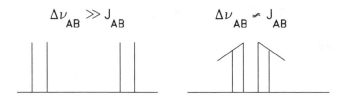

**Figure 2.14**   The splitting of the $AB$ proton system for $\Delta v_{AB} \cong J_{AB}$ (right) showing the "roof effect" and that for $\Delta v_{AB} >> J_{AB}$.

The difference between chemical and magnetic equivalence is sometimes misunderstood by chemists, as will be explained in the following example. Atomic nuclei in the same chemical environment, such as H1 and H2 protons (and F1 and F2 fluorine nuclei) in 1,1-difluoroethylene **54,** are *chemically equivalent.* They exhibit the same chemical shifts. However, the couplings between the protons and fluorine nuclei are different $J_{F1H1} \neq J_{F1H2}$ [60]. *Magnetically inequivalent* nuclei are denoted by primes (*AA'XX'* in the case of **54**), and their spectrum is more complicated than that of the $A_2X_2$ system. The latter can be exemplified by difluoromethane, which exhibits a triplet in the proton spectrum, while a complicated pattern consisting of at least 10 signals is observed in the proton spectrum of **54.**

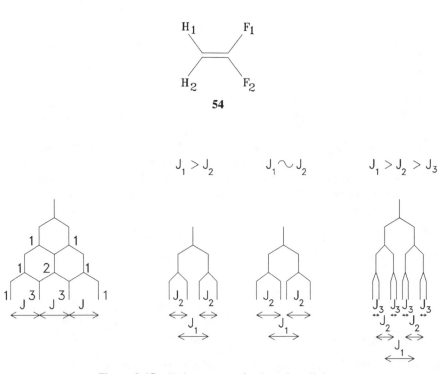

**Figure 2.15**   Various types of spin–spin splitting.

Coupling constants are very sensitive to the spatial arrangement of the respective nuclei. The most important in this respect is the dependence of the vicinal coupling constant through three consecutive bonds $^3J$ on the dihedral angle formed by these bonds, as is described by the Karplus relation [61]

$$^3J = A + B \cos \phi + C \cos 2\phi \qquad (2.28)$$

Different values of the $A$, $B$, $C$ constants have been proposed in the literature [62], but they all yield a bigger value of the constant for the *trans* arrangement than for the *cis* one. The dependence of vicinal coupling constants on other parameters (the length of the central bond, bond angles, electronegativity of substituents) is discussed in detail in the Günther [56] and Breitmaier [58] monographs. Similarly, the dependence of geminal coupling on the hybridization of the carbon and on the orientation of the neighboring $\pi$ bond or of nonbonding electron pairs is also briefly reviewed there.

### 2.3.8.2 Two-Dimensional NMR Experiments [56–58]

*Spin decoupling* (*double resonance*) consists in the simplest case of an *AX* system in the fact that the $J_{AX}$ splitting of the signal $A$ is removed by the irradiation of the second radio frequency corresponding to the Larmor frequency of the nucleus $X$. Then, as illustrated in Fig. 2.12(b) and (c), the signal $A$ will collapse to singlet, while in place of the signal $X$ an interference between the Larmor frequency of the nucleus $X$ and the decoupling frequency appears. If $A$ and $X$ are nuclei of the same kind, then one speaks about *selective homonuclear decoupling;* otherwise we deal with *heteronuclear decoupling.*

Three types of heteronuclear decoupling experiments, differing in the time of operation of the second irradiating signal, are important in $^{13}C$ spectra. (a) By broad-band heteronuclear proton decoupling, the coupling of all protons with the carbon atoms is simultaneously removed, yielding all the carbon signals as singlets. Figure 2.16 presents a $^{13}C$ spectrum of ethyl benzene **53** with (bottom) and without (top) decoupling. (b) Gated decoupling is a special type of broad-band decoupling experiment enabling the measurement of spectra with proton–carbon splittings enhanced by the nuclear Overhauser effect, to be discussed below. (c) By inverse gated decoupling the proton-carbon splitting is removed, and the carbon spectra reflect true quantitative relations among the singlet carbon signals. (As mentioned before, contrary to the intensities in proton spectra, the corresponding values in standard $^{13}C$ spectra cannot be used for quantitative determinations.) (d) Selective heteronuclear proton carbon decoupling is analogous to the selective homonuclear decoupling discussed previously. At present this method has been replaced by CH-COSY two-dimensional spectra, to be discussed later in this section.

An increase or decrease of the intensity of the NMR signals as a result of decoupling experiments is called the *nuclear Overhauser effect* (NOE) [63]. It enables one to detect which nuclei are close in space, although they are not directly bound. NOE studies are especially useful in supramolecular chemistry and in the investigation of proteins in solution. Its application is illustrated by selective irradia-

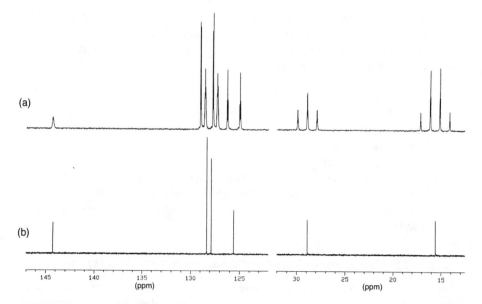

**Figure 2.16** (a) Decoupled and (b) fully coupled $^{13}C$ spectrum of ethyl benzene. (Courtesy of Mr. J. Sitkowski.)

**55**

tion of the H3, H5, and O2-, and O3-methyl protons of heptakis(2,6-di-O-Me)-β-cyclodextrin **36b** in the $^1H$ spectrum of its complex with diphenylporphyrin **55** [64]. The irradiation induces intensity changes in the signals of **36b** presented in Table 2.4. Together with information on the 1(porphyrin):2(cyclodextrins) stoichi-

**Table 2.4** NOEs in the 200-MHz $^1H$ Spectrum of Porphyrin **55** in a Complex with **36a**

| CD protons irradiated | Meso | $^1H_\alpha$ | $^1H_\beta$ | Ortho | Meta + para |
|---|---|---|---|---|---|
| $^2OMe$ | 16% | 0% | 0% | 0% | 0% |
| $^3H$ and $^5H$ | 0% | 0% | −25% | −1.2% | −11.7% |
| $^6OMe$ | 0% | 0% | 0% | 1.9% | −3.5% |

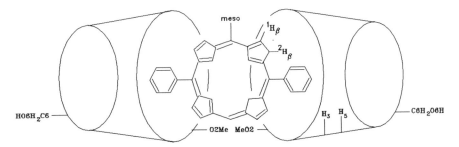

**Figure 2.17** Schematic representation of the 1:2 complex of diphenylporphyrin with heptakis(2,6-di-O-Me)-β-CyD.

ometry of the complex, the NOE information allowed the authors to propose the structure of the inclusion complex presented schematically in Fig. 2.17. Another example of a complex of hexakis(2,3,6-tri-O-methyl)-α-cyclodextrin **36a** with *para*-nitrophenol **37c** (X = NO$_2$) shows different structure in the solid state and in solution [14]. As shown schematically in Fig. 2.18, the mode of entrance of the guest **37c** into the host cavity in the solid state determined by X-ray studies [Fig. 2.18(a)] is the opposite of that determined in solution by NOE NMR experiments [Fig. 2.18(b)]. In the latter an irradiation of H5 protons of the host **36a** yielded an intensity increase of the signals of protons meta to the nitro group of the guest, while an irradiation of the H3 protons of **36a** produced an intensity increase of all aromatic protons of **37c**. However, as mentioned before, NOE can consist not only of an enhancement of the signals, but it can also produce (e.g., in case of the $^{15}$N nucleus) a decrease in their intensity. As a result, by an unfavorable parameter choice a signal can even disappear due to NOE effect.

The NOE effect can be also used to elucidate the spatial molecular structure in a quantitative way. An elegant example of its joint use with information stemming from spin–spin coupling constants and molecular mechanics calculations (discussed in Section 3.3) was published by Y. Yang and co-workers [65]. The authors determined the spatial structure of the natural compound nervosin **56** in solution. Their

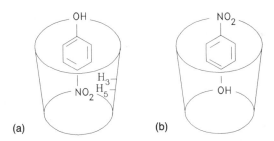

**Figure 2.18** Schematic representation of the mode of entrance of nitrophenol into the **36a** cavity in (a) the solid state, as determined by X-ray analysis and (b) in solution, as established by NOE [14].

**56**

study has shown that the solution structure of this polycyclic compound is close to that in the solid state as determined by the X-ray study [66].

As mentioned, by simple decoupling experiments a sample is irradiated with an additional electromagnetic frequency, usually corresponding to a resonance condition of a particular signal. By *two-* and *higher-dimensional spectra* the sample is irradiated by one or more additional carefully chosen pulse sequences, enabling one to reveal interrelations among the signals and, in turn, among absorbing magnetic nuclei. Correlation spectroscopy, abbreviated COSY, is the most frequently used type of two-dimensional NMR spectrum. In *homonuclear H,H-COSY* the resonance frequencies of coupled protons are correlated, showing the coupling pattern in a molecule under study. Thus in the H,H-COSY contour diagram of β-CyD **14b** presented in Fig. 2.19, the proton spectrum is recorded on both axes. The diagonal peaks correspond to the proton resonances, while the off-diagonal ones (correlations) exhibit the couplings. The latter enables an assignment of geminal and vicinal protons, as well as more distant protons in a zig-zag arrangement. Thanks to its effectiveness, homonuclear H,H-COSY has now replaced simple decoupling experiments.

Similarly to homonuclear COSY, *heteronuclear C,H-COSY* correlates $^1$H and $^{13}$C chemical shifts, which are displayed on the axes of the COSY diagram, and correlates the signals of directly bonded protons and carbon atoms. The connectivity among carbon atoms can be studied by means of another kind of two-dimensional experiment, *C,C–INADEQUATE*. The correlations between geminal and vicinal protons and carbon atoms are revealed by C,H-COLOC spectra, which enable an identification of quaternary carbon atoms nondistinguishable in C,H-COSY experiments.

### 2.3.8.3 Dynamic NMR Studies

NMR studies of dynamic phenomena [67] represent a major application of this technique. These may refer to internal dynamics of organic molecules (rotations around a single or partial double bond or ring inversion), inversion of configuration on a nitrogen atom, valence tautomerism, proton exchange, or more complicated chemical reactions. A process under investigation can be fast, slow, or of intermediate velocity on the NMR time scale. An interesting example of dynamic behavior is

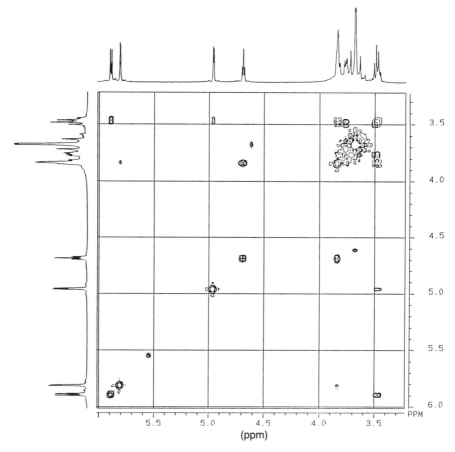

**Figure 2.19** Homonuclear COSY spectrum of β-CyD **14b**. (Courtesy of Mr. J. Sitkowski.)

provided by the spectra of the spirobisdioxane derivative **57c** [68], presented in Fig. 2.20. All six-membered rings in the molecule can undergo inversion. As discussed in Chapter 5, contrary to the Cahn, Ingold, and Prelog classification of chirality [69], spiro[5.5]undecane **57a,b** in frozen conformations is chiral, and its invertomers are enantiomers as well. Thus, having three spiro centers, **57c** exists as a mixture of three interconverting species fully extended (fe), extended-folded (ef), and fully folded (ff). The low-temperature dynamic spectra of the spirobisdioxane derivative **57c** presented in Fig. 2.20 exhibit separate signals of protons at C7, C8, C16, and C17 of all three species involved.

When a dynamic process is very fast, only the averaged signal appears characterized by the parameter value $\bar{P}$ averaged over the $P_A$ and $P_B$ describing the parameter for the nuclei $A$ and $B$:

$$\bar{P} = p_A P_A + p_B P_B \tag{2.29}$$

where $p_A$ and $p_B$ denote mole fractions of the components $A$ and $B$, respectively.

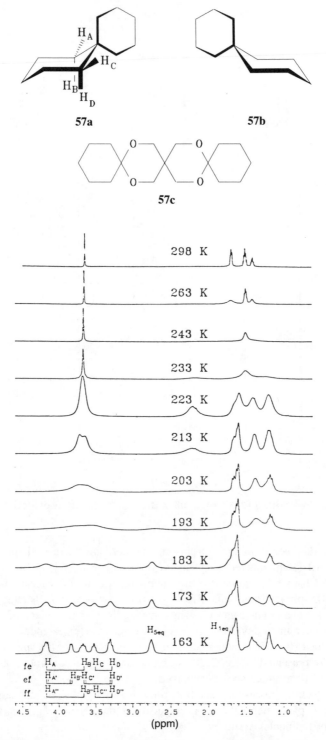

**Figure 2.20** Dynamic [^1]H spectra of spirobisdioxane derivative **57c.** (Courtesy of Mr. J. Sitkowski.)

Thus (see below) the averaged signal of inner and outer protons in [18]annulene **58** at 383 K lies at the average position of those signals calculated with 12:6 weights, corresponding to the population of the respective protons. Similarly, the observed value of the vicinal coupling constant $^3J$ of H5 and H6 protons in α-cyclodextrin **14a** is a weighted average of the corresponding values for two orientations resulting from internal rotation around the C5–C6 bond. The conformations with the third orientation (see formulae **59a–59c**) are not populated in this compound.

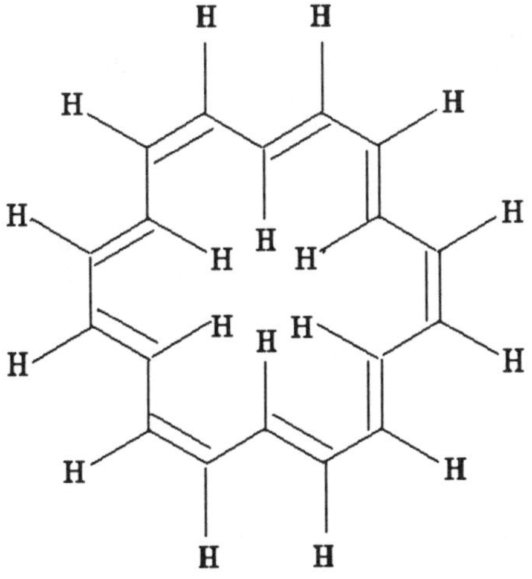

**58**

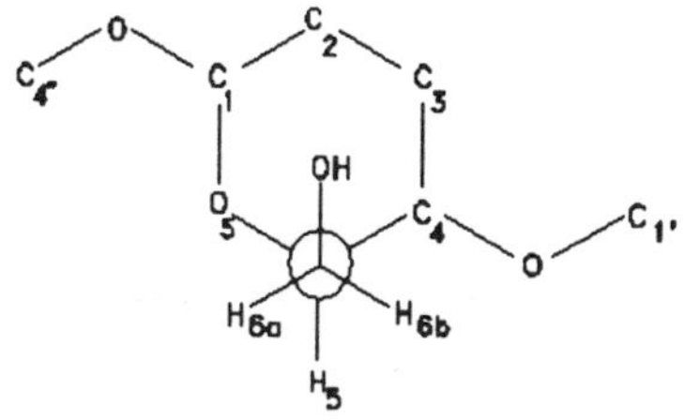

*gauche–gauche*

**59a**

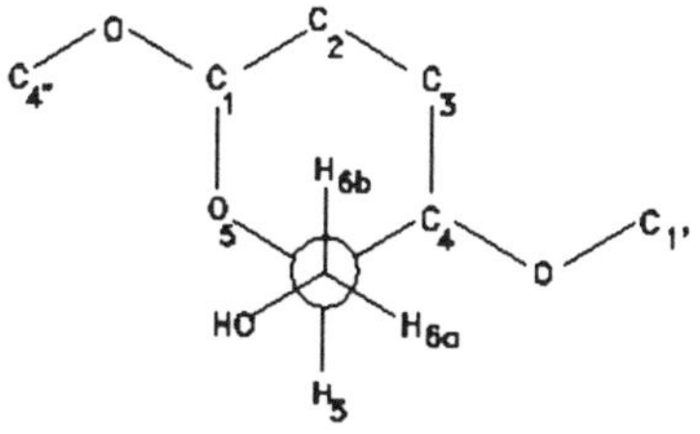

*gauche–trans*

**59b**

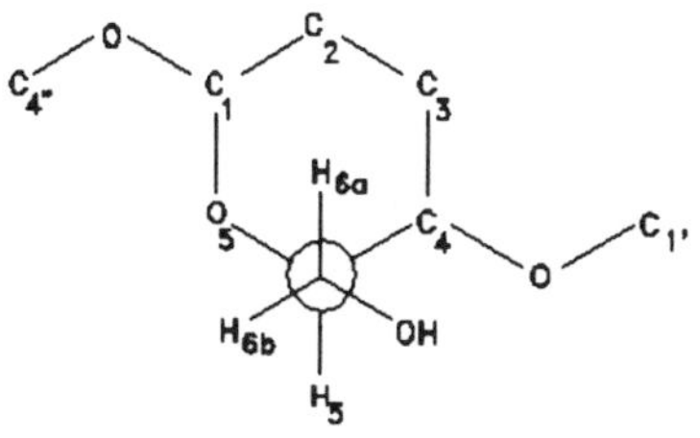

*trans–gauche*

**59c**

Dynamic processes with intermediate exchange rates are the most complicated to analyze. In many cases by cooling and/or heating the sample one can observe the full range of dynamic changes of the NMR spectrum, which can be simulated by calculations allowing one to determine the free enthalpy of activation of the process. NMR is applicable for the study of dynamic processes taking place with rates between $10^{-1}$ and $10^3$ s$^{-1}$. $^{13}$C NMR spectra of *cis*-decalin **25**, shown in Fig. 2.21, manifest such a process [70]. Contrary to the *trans*-isomer **26,** in which the type of ring junction prevents their inversion, at room temperature the molecule is known to exhibit a rapid inversion of cyclohexane rings, which is frozen at 223 K. Thus the 1, 4, 5, and 8 positions are equivalent at room temperature, while at low temperature

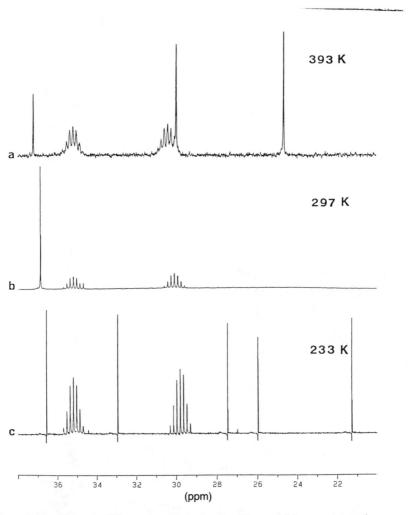

**Figure 2.21** Dynamic $^{13}$C spectra of *cis*-decalin. The multiplet structures shown correspond to traces of nondeuterated solvent. (Courtesy of Mr. J. Sitkowski.)

only the equivalence of positions 1 and 5 and that of positions 4 and 8 is retained. Thus at room temperature the signals of the 1 and 4 carbons exchange with those of the 5 and 8 carbons respectively, while the carbons C9 and C10 are equivalent and do not show dynamic behavior. The carbons at the 2, 3, 6, and 7 positions exhibit the same behavior. Therefore, the top spectrum shown in Fig. 2.21 measured at 333 K exhibits three signals with an approximate intensity ratio of 1:2:2, while in the spectrum measured at 223 K there are five signals with approximately equal intensities. Interestingly, at room temperature the dynamic signals of carbons C1 to C8 are at coalescence, which means that they disappear and the spectrum exhibits only one signal. If, as is usually the case, the first spectrum of *cis*-decalin is taken at room temperature, a first impression is slightly disappointing. A further complication of the spectrum arises when *cis*-decalin is inserted into the β-cyclodextrin cavity, since with the frozen ring inversion the guest molecule is chiral (it has $C_2$ symmetry; see the condition of chirality formulated in Chapter 5), and its invertomers **25a** and **25b** are enantiomers as well. In the absence of another chiral factor, they are indistinguishable by NMR. However, as will be shown in Chapter 10, the encapsulation of the *cis*-decalin invertomers into the β-cyclodextrin **14b** cavity produces additional splittings of the signals due to the formation of diastereomeric complexes by **25a** and **25b** with chiral **14b.**

Molecules having π electrons, and especially those possessing aromatic rings, exhibit considerable magnetic anisotropy, which can be useful for structure determination. In the latter case the system behaves like a loop, with electric current producing a secondary field that deshields protons in the molecular plane outside the ring, while protons below and above the plane are strongly shielded. The effect is particularly strong for the macrocyclic [18]annulene **58,** for which at $T = 213$ K δ (H outer) was found to be 9.28 ppm, while the corresponding value for the inner protons is equal to −2.99 ppm. As mentioned, at 373 K one averaged signal with δ = 5.4 ppm was observed close to the calculated average of 5.29 ppm [71]. An interesting example of a considerable magnetic anisotropy in a strongly deformed adamantane derivative **60** was reported by Vögtle and co-workers. Its X-ray study revealed that C11 is forced away from the aromatic ring, and the considerably deshielded protons $H'_i$ and $H''_i$ absorb at 294 K in toluene-$d_8$ **61** ($X = D$) at −0.07 and 0.27 ppm, respectively [72].

**60**     **61**

Interesting dynamic behavior has been also observed for synthetic oligomers such as **62** [73]. Some synthetic polymers are known to exhibit optical activity when prepared using helix-sense-selective polymerization. In solution their chiral helices can be rigid or racemize. To study such effects, **62** was synthesized, and its $^1$H NMR spectra measured at 303, 333, 345, and 378 K. Its room-temperature spectrum consisted of six relatively sharp signals corresponding to the five methine protons and two methyl groups in different environments. After going through coalescence, a single signal of the methyl protons appeared, while the signals of H1 and H2 methine protons were still broad. The dynamic process observed was interpreted as a change of sense of the helix and its rate-determining parameters were estimated from the spectra. At 293 K in toluene-$d_8$ **61** ($X = $ D) solution, the barrier to the change of helicity sense was found to be equal to 68.9 kJ/mol, and the rate of the process was determined to be 3.9 s$^{-1}$.

$$CH_3^a - O - \underset{CCl_3}{\overset{H^{1a}}{C}} - O - \underset{CCl_3}{\overset{H^{2a}}{C}} - O - \underset{CCl_3}{\overset{H^3}{C}} - O - \underset{CCl_3}{\overset{H^{2b}}{C}} - O - \underset{CCl_3}{\overset{H^{1b}}{C}} - O - CH_3^b$$

**62**

It is of primary importance that time scales of various spectroscopic methods are different, allowing an observation of various processes in the same molecule. *s-cis–s-trans* isomerism in enamino ketone **45** ($X = NR_2$, $R = CH_3$) is slow on an IR time scale and, in analogy with the spectra of vinyl ketones discussed earlier, the infrared spectrum is the sum of the bands corresponding to both conformers [74]. At room temperature the same process is fast on the NMR time scale, giving rise to the average spectrum of the mixture [75]. On the other hand, *cis–trans* isomerism on the double bond in substituted vinyl ketones **45** ($X = NR_2$, $R = CH_3$) is clearly manifested in room-temperature NMR spectra. Variable-temperature NMR spectra of both methylamino- and dimethylaminovinyl ketones allows one to determine the thermodynamic parameters of the process [76], which cannot be obtained by IR. This example once more shows the complementarity of different physical methods.

## 2.4 Scanning Probe Microscopy

Two different techniques, namely, *scanning tunneling microscopy* (STM) and *atomic force microscopy* (AFM), are encompassed in *scanning probe microscopy* (SPM) [77]. (The acronyms are applied both for the methods and the apparati used. In the latter cases M stands for the microscope.) They make use of an electric current flowing between two conductors separated by very small distances, of the order of angstroms. An extremely thin layer of vacuum, air, or liquid between the conductors acts as an insulator. The electrons "tunnel" (see Section 3.2.1 for a discussion of tunneling effect) through the layer, producing a measurable current exponentially dependent on the distance $d$ between the conductors, that is, on the thickness of the insulating layer. Let $I$ be the tunneling current; $V$ the voltage applied across the two conductors; and $\phi$, the effective tunneling barrier. Then

$$I = AV \exp(-c\phi^{-1/2})d \qquad (2.30)$$

where $A$ and $c$ are constants.

The operation schemes of STM and AFM is briefly presented. In SPM, the tunneling current flows between the probe configured as a tip to the surface under study. The probe plays the role of the first conductor, while the surface is the second. The methods critically depend on control over motion on a subangstrom scale executed by piezoelements [78]. This control allows the scanning of the sample, yielding a surface analysis with atomic resolution. This constitutes a major difference between SPM and the diffraction methods discussed in Section 2.2. X-ray and other diffraction techniques yield an image of the structure averaged over many atoms, while SPM provides direct local information atom by atom. Thus separate atoms on the face of a crystal lattice or in monolayers, as well as those in thin films such as Langmuir–Blodgett films (shortly presented in Section 10.4), can be directly measured and imaged. The differences between STM and AFM consist of (a) the resolution, since that provided by STM is of the order of 1 Å, thus enabling the observation of separate atoms, while that of AFM is much lower. Thus separate defects in the surfaces can be "seen" using STM while distinguishing different molecular orientations has only recently been achieved by AFM studies [77]. (b) The mechanism of operation of both techniques is not fully understood, but STM is known to be sensitive to the distance between the tip acting as the probe and the surface under study, while AFM reacts to repulsive or attractive interactions between the stylus and the surface. Magnetic force or electrostatic force microscopes (MFM and EFM, respectively) are specific types of AFM devices providing the possibility of imaging the corresponding properties of the surfaces under study.

SPM is a new analytical technique that is in a developmental stage. The equipment and experimental procedures are still in an initial stage of development and the operation principles are not fully understood. Therefore, their possibilities and limitations are still not settled. However, the method seems to offer, among others, unprecedented chances for studying inter- and intramolecular distances (the latter using STM). It will undoubtedly be useful in studies of organic conductors and semiconductors and tailored organic films, like Langmuir–Blodgett films or those formed by liquid crystals. It can also provide valuable technological information for microlithography and for the development of devices based on nonlinear optics (see Section 12.2.5). An example of an AFM image of a polymeric structure is presented in color chart 10.2.

This survey of physical methods as a source of information on the spatial structure of organic molecules does not purport completeness, but rather should compare the methods than go into details. A study of monographs on specific experimental techniques is highly recommended to the interested reader. Some other examples of the simultaneous application of various techniques for structure determination will be given in Section 7.2.4.3 and in Chapter 10.

## References

1a. J. D. Dunitz, *X-Ray Analysis and the Structure of Organic Molecules,* Cornell University Press, Ithaca, 1979, p. 310. b. pp. 325, 337, 363ff; c. A. I. Kiteygorodsky *Molecular Crystals and Molecules,* Academic Press, New York, 1973.

2. M. Dobler, J. D. Dunitz, B. Gubler, H. P. Weber, G. Büchi, and J. Padilla, *Proc. Chem. Soc.* (1963) 383.

3. G. Büchi, R. Erickson, and N. Wakabayashi, *J. Am. Chem. Soc.* **83** (1961) 927; G. Büchi and W. McLeod, *J. Am. Chem. Soc.* **84** (1962) 3205.

4. E. L. Mutterties, *Inorg. Chem.* **4** (1965) 769; A. Bassindale, in *Third Dimension in Organic Chemistry,* Wiley, New York, 1984.

5. P. W. Atkins, *Physical Chemistry,* Oxford University Press, Oxford, 1992.

6. J. M. Bijvoet, *Proc. Koninkl. Ned. Akad. Vetenschap.* **B52** (1949) 313.

7. J. M. Bijvoet, A. F. Peerdeman, and A. J. van Bommel, *Nature* (London), **168** (1951) 271; A. F. Peerdeman, A. J. van Bommel, and J. M. Bijvoet, *Proc. Koninkl. Ned. Akad. Vetenschap.* **B54** (1951) 16.

8. J. Trommel and J. M. Bijvoet, *Acta Cryst.* **7** (1954) 703.

9. See Chapter 5, in which the chirality of natural amino acids is briefly discussed.

10. H.-B. Bürgi, *Angew. Chem. Int. Ed. Engl.* **14** (1975) 460.

11a. F. H. Allen, S. Bellard, M. D. Brice, B. A. Cartwright, A. Doubleday, H. Higgs, T. Hummelink, B. G. Hummelink-Peters, O. Kennard, W. D. S. Motherwell, J. R. Rodgers, and D. G. Watson, *Acta Crystallogr. Sect. B: Struct. Crystallogr. Cryst. Chem.* **B35** (1979) 2331.

11b. Protein Data Bank, Chemistry Department, Building 555, Brookhaven National Laboratory, Upton, NY 11973.

11c. F. H. Allen, O. Kennard, D. G. Watson, L. Brammer, A. G. Orpen, and R. Taylor, *J. Chem. Soc., Perkin Trans. II* (1987) S1.

11d. P. Gilli, V. Bertolasi, V. Ferretti, and G. Gilli, *J. Am. Chem. Soc.* **116** (1994) 909.

12. O. Bastiansen, *Acta Chem. Scand.* **6** (1952) 205; C. P. Brock and K. L. Haller, *J. Phys. Chem.* **88** (1984) 3570; G. P. Charbonneau and Y. Delugeard, *Acta Ctrystallogr., Sect. B* **32** (1976) 1420.

13. J. D. Dunitz, *Pure Appl. Chem.* **63** (1991) 177.

14. Y. Inoue, Y. Takahashi, and R. Chujo, *Carbohydr. Res.* **114** (1985) C9.

15. W. C. Hamilton and S. J. La Placa, *Acta Cryst.* **B24** (1968) 1147.

16. W. C. Hamilton, J. W. Edmonds, and A. Tippe, *Disc. Farad. Soc.* **48** (1969) 192.

17a. K. Lindner and W. Saenger, *Carbohydr. Res.* **99** (1982) 103; T. Fujiwara, M. Yamazaki, Y. Tomiza, R. Tokuoka, K. Tomita, T. Matsuo, H. Suga, and W. Saenger, *Nippon Kagaku Kaishi* **2** (1983) 181.

17b. C. Betzel, W. Saenger, B. E. Hingerty, and G. M. Brown, *J. Am. Chem. Soc.* **106** (1984) 7545; V. Zabel, W. Saenger, and S. A. Mason, *J. Am. Chem. Soc.* **108** (1986) 3664;

17c. J. E. H. Köhler, W. Saenger, and W. F. Van Gunsteren, *Eur. J. Biophys.* **15** (1987) 211;

17d. B. Lesyng and W. Saenger, *Biochim. Biophys. Acta* **678** (1981) 408.

18. P. Gluzinski, J. W. Krajewski, B. Korybut-Daszkiewicz, A. Mishnyov, and A. Kemme, *J. Cryst. Spectr. Res.* **23** (1993) 61.

19. C. S. Yannoni, P. P. Bernier, D. S. Bethune, G. Meijer, and J. R. Salem, *J. Am. Chem. Soc.* **133** (1991) 3205.

20. N. E. Heimer and D. L. Mattern, *J. Am. Chem. Soc.* **115** (1993) 2217.

21. G. E. Bacon, *Neutron Diffraction,* Oxford University Press, Oxford, 1975.

22. L. V. Vilkov, V. S. Mastryukov, and N. I. Sadova, *Determination of the Geometrical Structure of Free Molecules,* Mir Publishers, Moscow, 1983.

23. I. Hargittai and G. Schultz, *J. Chem. Soc., Chem. Commun.* (1972) 323.

24. V. S. Mastryukov, O. V. Dorofeeva, L. V. Vilkov, and I. Hargittai, *J. Mol. Struct.* **34** (1976) 99.

25. R. L. Hilderbrandt, J. D. Wieser, and L. K. Montgomery, *J. Am. Chem. Soc.* **95** (1973) 8598.

26. K. Hedberg, in *Critical Evaluation of Chemical and Physical Structural Information,* D. R. Lide, Jr., and M. A. Paul, Eds., National Academy of Sciences, 1974, p. 77.

27. V. P. Novikov, L. S. Khaikin, L. V. Vilkov, et al., *Zh. Obshch. Khim.* **47** (1977) 958.

28. M. Hargittai and I. Hargittai, "The Importance of Small Structural Differences," in *Molecular Structure and Energetics,* J. F. Liebman, and A. Greenberg, Eds., Vol. 2, p. 1.

29. J. Konarski, *J. Mol. Struct.* **275** (1992) 1; M. Molski, *J. Konarski Phys. Rev.* **A47** (1993) 711.

30. W. Gordy and R. L. Cook, *Microwave Molecular Spectra,* Wiley, New York, 1984.

31. N. W. Broten, T. Oka, L. W. Avery, J. M. Macleod, and H. W. Kroto, *Astrophys. J.* **223** (1978) L105.

32. I. Ozier, *Phys. Rev. Lett.* **27** (1971) 1329; P. W. Atkins and J. A. M. F. Gomes, *Chem. Phys. Lett.* **39** (1976) 519.

33. C. E. Cleeton and N. H. Williams, *Phys. Rev.* **45** (1934) 234.

34. S. J. Cyvin, Ed., *Molecular Structures and Vibrations,* Elsevier, Amsterdam, 1972, p. 533.

35. L. Hedberg, K. Hedberg, Ph. E. Eaton, N. Nodari, and A. G. Robiette, *J. Am. Chem. Soc.* **113** (1991) 1514.

36. G. Herzberg, *Infrared and Raman Spectra of Polyatomic Molecules,* Van Nostrand, New York, 1954;

37. L. J. Bellamy, *The Infrared Spectra of Complex Molecules,* Chapman and Hall, London, 1980.

38. K. Nakanishi, *Infrared Absorption Spectroscopy. Practical,* Holden-Day, Inc., San Francisco, 1972.

39. C. G. Le Fevre, R. J. W. Le Fevre, R. Roper, and R. K. Pierens, *Proc. Chem. Soc.* (1960) 117.

40. E. B. Wilson, Jr., J. C. Decius, and P. C. Cross, *Molecular Vibrations. The Theory of Infrared and Raman Vibrational Spectra,* McGraw-Hill, London, 1955, pp. 367, 370.

41. Ref. 40, p. 244.

42. J. Kossanyi, *Bull. Soc. Chim. Fr.* (1965) 704.

43. L. P. Kuhn, *J. Am. Chem. Soc.* **80** (1958) 5950.

44. W. T. Aleksanian, G. M. Kuz'yantz, M. Yu. Lukina, S. V. Zotova, and E. I. Vostokova, *Zh. Strukt. Khim.* **9** (1968) 141.

45. M. J. Almond and A. J. Downs, *Spectroscopy of Matrix Isolated Species,* Wiley, New York, 1989; *Chemistry and Physics of Matrix-Isolated Species,* L. Anfrews and M. Moskovits, Eds., North-Holland, Amsterdam, 1989.

46. J. G. Radziszewski, T.-K. Yin, G. E. Renzoni, D. A. Hrovat, W. T. Borden, and J. Michl, *J. Am. Chem. Soc.* **115** (1993) 1454.

47. G. Maier, *Chem. Ber.* **125** (1992) 265.

48. Ref. 40, p. 244.

49. G. Herzberg, *Electronic Spectra and Electronic Structure of Polyatomic Molecules,* Van Nostrand, New York, 1966.

50. S. F. Mason, *Molecular Optical Activity and the Chiral Discriminations,* Cambridge University Press, Cambridge, 1982.

51. G. Snatzke, *Chemie Unserer Zeit* (in German) **15** (1981) 78, **16** (1982) 160.

52. M. Legrand and M. J. Rougier, "Application of the Optical Activity to Stereochemical Determinations," in *Stereochemistry,* H. B. Kagan, Ed., Vol. 2, p. 33, Thieme-Verlag, Stuttgart, 1977.

53. T. Polonski, M. J. Milewska, and A. Katrusiak, *J. Am. Chem. Soc.* **115** (1993) 11410.

54. N. J. Greenfield and G. D. Fasman, *Biochem.* **8** (1969) 4108.

55. G. Patonay and J. M. Warner, *J. Incl. Phenom. Mol. Recogn. Chem.* **11** (1991) 313.

56. H. Günther, *NMR Spectroscopy,* J. Wiley & Sons, Chichester, 1980; (a) p. 39. (b) pp. 107–13.

57. J. K. M. Sanders and B. K. Hunter, *Modern NMR Spectroscopy,* Oxford University Press, Oxford, 1993.

58. E. Breitmaier, *Vom NMR Spectrum zur Strukturformel Organischer Verbindungen,* B. G. Teubner, Stuttgart, 1992.

59. E. Breitmaier and W. Völter, *Carbon-13 NMR Spectroscopy–High Resolution Methods and Applications in Organic Chemistry and Biochemistry,* VCH, Weinheim, 1990.

60. E. D. Becker, *High Resolution NMR,* Academic Press, New York, 1969, p. 310.

61. M. Karplus, *J. Chem. Phys.* **30** (1959) 11.

62. K. Imai and E. Osawa, *Magn. Res. Chem.* **28** (1990) 668.

63. D. Neuhaus and M. Williamson, *The Nuclear Overhauser Effect in Structural and Conformational Analysis,* VCH Publishers, New York, 1989.

64. J. S. Manka and D. S. Lawrance, *Tetrahedr. Lett.* **30** (1989) 7341.

65. Y. Yang, X. Xu, D. Wang, and B. Qian, *Magn. Res. Chem.* **32** (1994) 71.

66. J.-K. Zhai, W.-C. Han, X.-H. Ju, W. T. Yu, and Q.-Z. Zhao, Communication at the International Symposium on Modern Chemistry, Zhengzhou, China, 1990, cited in Ref. 62a.

67. M. Oki, *The Chemistry of Rotational Isomers,* Springer, Berlin, 1993, p. 51.

68. H. Dodziuk, J. Sitkowski, L. Stefaniak, I. G. Mursakulov, I. G. Gasanov, and V. A. Kurbanova, *Struct. Chem.* **3** (1992) 269.

69. R. S. Cahn, C. K. Ingold, and V. Prelog, *Angew. Chem., Int. Ed. Engl.* **5** (1966) 385.

70. H. Dodziuk, J. Sitkowski, L. Stefaniak, J. Jurczak, and D. Sybilska, *J. Chem. Soc., Chem. Commun.* (1992) 207.

71. M. Gossman and B. Franck, *Angew. Chem.* **98** (1986) 1107; G. Kübel and B. Franck, *ibid.* **100** (1988) 1203;

72. F. Vögtle, J. Dohm, and K. Rissanen, *Angew. Chem., Int. Ed. Engl.* **29** (1990) 902.

73. K. Ute, K. Hirose, H. Kashimoto, and K. Hatade, *J. Am. Chem. Soc.* **113** (1991) 6305.

74. H. Dodziuk, K. Kamieńska-Trela, and J. Da₋browski, *Ann. Soc. Chim. Polon.* **44** (1970) 393.

75. J. Da₋browski and L. Kozerski, *J. Chem. Soc. Chem. Commun.* (1968) 586.

76. J. Da₋browski and L. Kozerski, *J. Chem. Soc. B* (1971) 345.

77. J. Frommer, *Angew. Chem., Int. Ed. Engl.* **31** (1992) 1298.

78. R. Overney, L. Howald, J. Frommer, E. Meyer, and J. Gunterodt, *J. Chem. Phys.* **94** (1991) 8441; M. Egger, F. Ohnesorge, A. Weisenhorn, S. Heyn, B. Drake, C. Prater, S. Gould, P. Hansma, and H. Gaub, *J. Struct. Biol.* **103** (1990) 89.

# 3

# Dry Chemistry: Theoretical Calculations as a Powerful Method of Structure Determination

## 3.1 Introduction

"Dry chemistry" (i.e., theoretical calculations) is playing an ever-increasing role in the structure elucidation of organic molecules. Moreover, novel systems are being proposed as plausible synthetic targets on the basis of theoretical investigations. The rapid development of computers and of accompanying computer graphics and user-friendly programs has made it possible gradually to bridge the gap between synthetic and theoretical chemists. On one hand, modern powerful computers enable one to carry out reliable calculations for systems that were not amenable to theoretical treatment a few years ago. On the other hand, PCs have become standard equipment in synthetic laboratories, where they are not only used as intelligent typewriters. More powerful workstations with amazing graphic capabilities have become widely available. The ease of the applying modern programs and the aesthetic appeal of the graphical images produced have attracted numerous nonspecialists to the field. The images are beautiful and appealing. Moreover, user-friendly programs create an impression that theoretical chemistry is simple. As a consequence, the complexity of theoretical methods and the problem of the reliability of the results is not fully recognized by those who perform the calculations. We believe that these difficulties can be overcome only by cooperation between theoretical and organic chemists.

Applications of theoretical chemistry can be approximately divided into three main groups: quantum chemistry calculations (QC) [1], molecular mechanics (MM) [2], and molecular dynamics (MD) [3]. MM is sometimes called the force-field (FF) method, while for the latter the term dynamic simulations (SD) is also applied [4]. For larger molecules hybrid methods combining QC and MM calculations are often used. Today (that is, in 1994) MM calculations can be carried out for molecules consisting of fewer than 20,000 atoms. Molecules having 200–1000 atoms are tractable by semiempirical QC. MD yields reliable results for systems consisting of ca. 100 atoms. 100-atom systems appear to be the limit for typical Hartree–Fock QC, while ab initio calculations taking into account correlation effects are usually performed for systems having fewer than ca. 20 atoms. (These figures are given only to show the order of magnitude of the systems that can be reliably studied theoretically. With the continuing rapid development of computers, they will certainly become outdated by the time this book will have been printed.)

It should be stressed that only ab initio QCs have a sound theoretical basis. MM and MD methods are based on empirical model potentials, with very little theoretical justification. Nevertheless, on the basis of numerous results, it is well established that MM calculations yield very reliable and accurate data, especially those concerning the spatial structure of hydrocarbons. Until recently, reliable QCs could be carried out only for very small molecules. This situation has drastically changed. Today fairly accurate calculations can be carried out for medium-sized molecules of interest to synthetic chemists. This enables not only a rationalization of experimental trends, but also predictions based on studies of hypothetical molecules or of systems that are impossible or difficult to be studied experimentally.

As mentioned, one of the most exciting aspects of such studies is the possibility of proposing new hypothetical molecules with interesting properties as plausible synthetic targets. Sometimes, as was the case with $C_{60}$ **11** discussed in Section 8.5, such predictions are premature with respect to experiments [5, 6], and they do not influence synthetic work. However, the impact of well-timed calculations on experimental studies can be very strong. As described in Section 7.2.2, calculations of the energy required to planarize methane yielded very different values [7]. Nevertheless, they inspired massive synthetic efforts and resulted in the syntheses of numerous fenestranes **63** [8]. Paradoxically, neither the known fenestranes nor hypothetical windowpane **63a,** which was thought to have a planar central carbon atom, can possess such a configuration. Thus, as described in Section 7.2.2, in the case of "the planar methane problem," calculations have not led to the synthesis of a desired molecule, but they did stimulate many interesting syntheses. However, the existence of small-ring propellanes **64** having inverted carbon atoms was predicted on the basis of the QC calculations by Wiberg [9], who later synthesized them. In the latter group of compounds, particularly surprising was the predicted greater stability

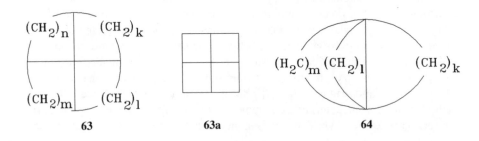

63                                    63a                                    64

of [1.1.1]propellane **5** (Section 7.2.1) than those of higher propellanes. The Wiberg synthesis of **5** and its studies confirmed his earlier theoretical predictions. At the same time they inspired the formulation of several questions concerning the properties or even the existence of the central, interbridgehead bond in [1.1.1]propellane. These, in turn, have led to many additional calculations, allowing one to propose a QC definition of the chemical bond. These problems are discussed in Section 7.2.1 in some detail.

As will be shown in this chapter, quantum chemistry calculations have a sound theoretical basis, but a molecule (or a system of molecules) is treated as a whole in such calculations. Therefore, some comments on the correspondence between the results of the calculations and such useful chemical concepts as bond length, aromaticity, $\pi$ stacking, or steric interactions will be discussed in Section 3.2.5 after a brief presentation of the foundations of QC. Contrary to ab initio quantum chemistry, MM and MD calculations, which will be briefly presented later, have a much weaker theoretical basis. Their use receives a posteriori justification by the astonishing accuracy of the results of these calculations. Their additional merit consists in their model character, allowing one to make predictions concerning analogous systems.

## 3.2 Quantum Chemistry Calculations

### 3.2.1 The Foundations

It became clear at the end of the nineteenth century that atomic spectra, blackbody radiation, and the photoelectric effect, could not be reconciled with classical mechanics. To explain these and other phenomena, quantum mechanics was formulated on the basis of some far-reaching postulates and principles. In the following a brief outline of the principles of nonrelativistic QCs [1, 9] will be presented. For a more precise and complete description of the ideas and concepts of quantum mechanics, in particular, as applied in quantum chemistry, the interested reader is referred to Ref. 1. A more simplified treatment of the same topic is presented in Refs. 10 and 11.

In quantum chemistry the state of a system is described by a wave function $\Psi(q_1, q_2, q_3, \ldots, t)$ of the coordinates of all particles forming the system and time. $|\Psi|^2 \, d\tau$ describes the probability of finding a particle in the space element $d\tau$. Thus $\Psi$ has to be continuous and can assume only finite values. The value of the integral $\int |\Psi|^2 d\tau$ also has to be finite since it describes the probability of finding the particle somewhere in space.

*Observables* are represented mathematically by operators. An operator corresponds to each variable (coordinate, momentum, energy, etc.), and functional relations characteristic of classical mechanics are replaced by the corresponding relations among operators. Thus the operator describing a coordinate of a particle acting on the function $f$ simply corresponds to the multiplication of this function by this coordinate $xf = xf$. Similarly, momentum operators assume the form

$p_x = -(ih/2\pi) \, \delta/\delta x$, $p_y = -(ih/2\pi) \, \delta/\delta y$, $p_z = -(ih/2\pi) \, \delta/\delta z$, where $i$ is the imaginary number. By classical analogy the full energy of a system is equal to the sum of its potential and kinetic energy $E = T + V$. Thus the Hamiltonian $H$ assumes the form (in atomic units [12], as used throughout this chapter)

$$H = T + V = (p_x^2 + p_y^2 + p_z^2) + V(q, t)$$

$$= -\frac{1}{2m}\left(\frac{\delta^2}{\delta x^2} + \frac{\delta^2}{\delta y^2} + \frac{\delta^2}{\delta z^2}\right) + V(q, t) \tag{3.1}$$

The wave function should satisfy the Schrödinger equation:

$$H(p, q, t)\Psi(q, t) = i\frac{\delta}{\delta t}\,\Psi(q, t) \tag{3.2}$$

Only stationary states, that is, the states that are not time dependent, are considered in conformational analysis. In this case, the Hamiltonian is not a function of time. Then a time-independent form of the Schrödinger equation can be applied

$$H\Psi(q) = E\Psi(q) \tag{3.3}$$

The important point is that this equation has physically admissible solutions only for certain values of the energy $E$, each characterizing a stationary state. Thus the energy and other observables are quantized.

For a hydrogenlike atom, expressing Eq. (3.3) in polar coordinates leads to three equations for $r, \vartheta, \phi$ that can be solved only for integer values of the constants $n, l, m$ called *quantum numbers*. The number $n$ assumes zero and positive values, $n > 1 \geq 0$, while $m$ is equal to $0, \pm 1, \pm 2, \ldots, \pm 1$. Traditionally, the wave functions (that is, the solution of Eq. (3.3) called orbitals) with $l = 0$ are denoted by the letter $s$, those with $l = 1$ by the letter $p$, while $d, f, g$, and the following letters in the alphabet are used for $l = 2, 3, 4$, etc., respectively. The functions differ in their symmetry properties. The totally symmetric $s$ functions are represented by spheres of radii increasing with $n$. There are three $p$ functions, denoted $p_x, p_y$, and $p_z$, with $p_x$ antisymmetric with respect to reflections in the $x = 0$ plane and symmetric with respect to reflections in the $y = 0$ and $z = 0$ planes, etc.

Similar results are obtained for many-electron atoms. For the ground state the energy levels $E_{nl}$ corresponding to a given set of quantum numbers should be occupied in the order of their increase. (In a first approximation the energy values corresponding to different $m$ values are degenerate; that is, $E_{nl}$ values do not depend on $m$.) There is, however, a limit to the occupancy of the energy levels. The *Pauli exclusion principle* states that the orbital corresponding to the specific values of $n, l, m$, quantum numbers can be occupied by only two electrons. The classical quantities (coordinates, momenta, and energy) were shown to be insufficient to explain atomic and molecular structure. Thus the operators of spin momentum $s$ and its projection on the $z$ axis $s_z$ had to be introduced. These quantities do not have any classical analogy, and they cannot be expressed as functions of the particle coordinates and momenta [13]. The spin quantum number of electrons $s$ assumes only the $\frac{1}{2}$ value.

To introduce spin into the Schrödinger equation, the wave function $\Psi$ must be a function of both spin $\omega$ and spatial $q$ coordinates. Spin can adopt two "orientations," up and down, corresponding to the projection of spin on the $z$ axis $s_z = \pm\frac{1}{2}$. Their corresponding wave functions are usually denoted $\alpha(\omega)$ and $\beta(\omega)$, respectively. It should be stressed that the orientations are defined only in the presence of an additional factor, such as a magnetic field, which imposes a definite direction in space. For one electron

$$\Psi(x) = \Psi(q, \omega) = \phi(q)\alpha(\omega) \text{ or } \phi(q)\beta(\omega) \tag{3.4}$$

The Schrödinger equation (3.3) can be solved exactly only for a few simple cases. For most molecules approximate methods have to be applied to find solutions. These are often based on the variational or perturbational approaches. The former are based on the *variation principle,* stating that for a given approximate wave function $\psi'$ the expectation value of the Hamiltonian $\int\psi'^*H\psi'd\tau$ is an upper bound to the exact ground-state energy. This means that the energy of the system calculated using an approximate wave function is always higher than the accurate value. An approximation that yields lower energy is considered to be better. The *perturbation methods* are applied when an interaction in the system under study can be treated as a small perturbation of the system. The solution of the Schrödinger equation (3.5) for the unperturbed system is assumed to be known.

$$H^0\Psi^0(q) = E^0\Psi^0(q) \tag{3.5}$$

The perturbed Hamiltonian and its wave functions are assumed in the form of a power series as functions of a small perturbation parameter $\lambda$

$$H = H^{(0)} + \lambda H^{(1)} + \lambda^2 H^{(2)} + \cdots$$
$$\psi = \psi^{(0)} + \lambda\psi^{(1)} + \lambda^2\psi^{(2)} + \cdots$$

$\lambda$ is assumed to be small. Thus the preceding series are truncated. In first-order perturbation theory only terms with $\lambda$ are included. In second order the terms with $\lambda^2$ are also included, and so on. Substituting the expansions into Eq. (3.5), one calculates the values of $H^{(1)}$, $H^{(2)}$, etc., the corresponding corrections to the energy $E^0$. Similarly, $\psi^{(1)}$, $\psi^{(2)}$, etc. are the corresponding corrections to the unperturbed wave function $\psi^0$. For example, calculations of spectroscopic properties may be carried out treating the molecule–radiation interaction as a small perturbation.

An important quantum phenomenon that has no classical analogy is the *tunneling effect.* If one considers a particle passing through an energy barrier $V_0$, then classically it can go through this barrier only if its energy is larger than $V_0$. Otherwise this particle will stay in an energy well. In quantum mechanics there is a finite probability of the barrier passage even when the particle's energy is lower than the barrier $V_0$.

Approximate solutions of the Schrödinger equation for a many-electron atom can be obtained in the form of one-electron wave functions called *orbitals* and the orbital energy levels. These can also be characterized by quantum numbers $n$, $l$, $m$, and $s$. The functions depending on spatial and spin coordinates are called *spin orbitals,* while those depending on spatial coordinates only are called *spatial orbit-*

*als* or simply *orbitals*. In the ground state the orbitals are occupied from the lowest-energy one in increasing energy order and according to the Pauli principle. If a level is fully occupied, then a closed shell is formed. The highest occupied level is called the *valence shell*. The electrons occupying it are the valence electrons. A state with an electron promoted from an occupied level to an unoccupied level is called an *excited state*. Such a transition gives rise to a band in the electronic spectrum.

The lowest orbital with $n = 0$ may be occupied by at most two electrons. This configuration of the helium atom is denoted by $1s^2$. Similarly, for $n = 1$ and $l = 0$, there is a $2s^1$ level for the Li atom and the doubly occupied $2s^2$ level for the beryllium atom. As described earlier, the three levels for $n = p = 1$ are denoted $p_x$, $p_y$, and $p_z$. Thus the *electron configurations* (not to be confused with the geometrical one describing the arrangement of atomic nuclei in space discussed in Chapter 1) of the ground states of carbon, nitrogen, and oxygen atoms are denoted $1s^2 2s^2 2p^2$, $1s^2 2s^2 2p^3$, and $1s^2 2s^2 2p^4$, respectively. In organic compounds the electron configuration of carbon atoms sometimes may be better described involving the so-called valence-state configuration obtained by mixing the $2s$ and $2p$ orbitals. The mixing is especially effective for the carbon atom, for which the energy difference of these two orbitals is small. Three ways of mixing the $2s$ and $2p$ orbitals are most frequently used. The first one corresponding to the bonding in methane yields four equivalent orbitals in a tetrahedral arrangement. The resulting orbital mixing, called *hybridization*, is denoted $sp^3$. The next, better suited to describe the bonding in ethylene, consists of a $2p_z$ orbital and three equivalent orbitals formed from one $2s$ and two $2p_x$ and $2p_y$ orbitals. The latter defines $sp^2$ hybridization on carbon. This hybridization is characterized by the carbon atoms lying in a plane with their substituents. $\sigma$ bonds formed by $sp^2$ orbitals lie in the same plane, while the $\pi$ bond formed by $2p_z$ orbitals of the carbon atoms is perpendicular to this plane. For carbon there is also the possibility of mixing of one $2s$ and one $2p_x$ orbital. Then, as is the case in acetylene, these $sp$ orbitals on two carbon atoms form a $\sigma$ bond, while the corresponding $2p_y$, $2p_z$ orbitals form two perpendicular $\pi$ bonds.

## 3.2.2 Born–Oppenheimer Approximation

Unfortunately, the Schrödinger equation can be solved exactly only for the hydrogen atom and the $H_2^+$ molecular ion. Several approximations have to be made for solutions for more complicated systems. The most common is the Born–Oppenheimer approximation, making use, roughly speaking, of the fact that the movement of nuclei is much slower than that of electrons. In this approximation the wave function of a molecule $\phi$ is assumed to be the product of the corresponding electronic wave function $\phi_e$, vibrational wave function $\phi_\vartheta$, and rotational wave function $\phi_\varphi$:

$$\phi = \phi_e^* \phi_\vartheta^* \phi_\varphi \tag{3.6}$$

and the energy of the molecule described by the wave function $\phi$ is equal to the sum of independent terms characterizing electronic, vibrational, and rotational energy. The translational movement of the molecule as a whole is not quantized. Thus it does not produce any interesting effects on the molecular structure, and it is not analyzed in this chapter.

$$E = E_e + E_\vartheta + E_\varphi \tag{3.7}$$

The considerable energy difference between the preceding energy terms justifies a separate study of variables describing electronic, vibrational, and rotational transitions. As said before, the separation of the variables is made possible by the fact that the movement of electrons is much faster than that of the nuclei. Thus one can analyze electron movement in the field of (quasi)static nuclei. On the other hand, the movement of the latter is executed in the average field of electrons. Then electronic, vibrational, and rotational eigenproblems described by Eqs. (3.8)–(3.10), corresponding to the absorption of radiation in microwave, infrared, and ultraviolet–visible regions, respectively, can be solved independently. However, it should be stressed that the equations are coupled, and the electronic one (3.8) contains nuclear coordinates as parameters.

$$H_e \phi_e = E_e \phi_e \tag{3.8}$$

$$H_\vartheta \phi_\vartheta = E_\vartheta \phi_\vartheta \tag{3.9}$$

$$H_\varphi \phi_\varphi = E_\varphi \phi_\varphi \tag{3.10}$$

Electronic, vibrational, and rotational energy levels of a molecule are schematically presented in Fig. 2.7.

### 3.2.3 Entropy and Other Thermodynamic Functions

At absolute zero temperature only the lowest level is populated. At higher temperatures, higher rotational and vibrational levels are also populated. To describe molecular energy at temperatures higher than absolute zero, statistical sum, $Z$, is introduced [14]:

$$Z = \sum_i g_i \exp(-E_i/kT) \tag{3.11}$$

where $E_i$ denotes the energy levels of the system, $k$ is the Boltzmann constant, $T$ is absolute temperature of the system, and $g_i$ are the degeneracy factors of the $i$th level. The entropy $S$ is a measure of the diversity of the energy distribution over the levels. The entropy change is expressed as a logarithmic function of the statistical sum $Z$ [see Eq. (3.15)].

The most important thermodynamic functions of the ideal gas (i.e., free energy $F$, internal energy $E$, enthalpy $H$, and entropy $S$) can be expressed in the following way for one mole:

$$F = E_o - RT \ln Z - RT[\ln(2\pi mkT)^{3/2}V/(h^3N_A) + 1] \tag{3.12}$$

$$E = E_o + RT^2 \frac{d \ln Z}{dT} + \tfrac{3}{2}RT \tag{3.13}$$

$$H = E_o + RT^2 \frac{d \ln Z}{dT} + \tfrac{5}{2}RT \tag{3.14}$$

$$S = R \ln Z + RT \frac{d \ln Z}{dT} + R[\ln(2\pi mkT)^{3/2}V/(h^3N_A) + \tfrac{5}{2}RT \tag{3.15}$$

where increments from translational motion are included in square brackets, $R$ is the gas constant, $Z$ denotes the internal degrees of freedom (thus it is $Z_{intern}$), $E_0$ is the energy at absolute zero temperature, and $N_A$ is Avogadro's number.

It can be shown that the statistical sum, in turn, is equal to the product of the statistical sum related to the internal energy of the molecule and those associated with its translational and rotational motion. The first term can be expressed as the product of statistical sums associated with molecular vibrations and internal rotations. The translational term can be calculated on the basis of classical physics, and those associated with vibrations and internal rotations are determined by solving the Schrödinger equation in the Born–Oppenheimer approximation. Comparison of the experimental value of entropy or other thermodynamic functions with the value calculated for a given molecule with an assumed value of the barrier height for internal rotation allows one to estimate the barrier. This was exactly the case with the barrier to internal rotation in ethane when the difference between the calculated and experimental values forced Pitzer to assume a hindering of the rotation around the C—C bond and to determine the barrier to this rotation [15]. Until that time the rotation around a single C—C bond was considered to be completely free.

Entropy calculations can only be carried out for very small molecules. In many cases only contributions from the entropy of mixing and symmetry, to be discussed, are taken into account. In most cases entropy contributions for larger molecules are neglected. Reliable (see below) MD calculations allow one to estimate entropy changes in the system under investigation [16].

The relative populations of the $i$ and $j$ levels as a function of $\Delta E$, the energy difference between them, is expressed by the Boltzmann equation:

$$n_i/n_j = (g_i/g_j) \exp(-\Delta E/kT) = (g_i/g_j) \exp(-h\nu/kT) \tag{3.16}$$

where $n_i$, $n_j$ denote the number of molecules in the $i$th and $j$th level, respectively, $g_i$, $g_j$ denote degeneracies of the levels, $k$ is the Boltzmann constant, and $T$ the absolute temperature.

It can be shown that for a mixture of $N$ conformers characterized by molar fractions $N_i$ the entropy of mixing is equal to

$$\Delta S = R \sum_i N_i \ln N_i \tag{3.17}$$

where $N = \Sigma N_i$. Similarly, the entropy contribution associated with molecular symmetry is expressed as $\Delta S = -R \ln \sigma$, where $\sigma$ is the *symmetry order* [17], the number of times one arrives at equivalent structures by rotating the molecule around a symmetry axis.

### 3.2.4 Solving the Molecular Schrödinger Equation

Assuming the validity of the Born–Oppenheimer approximation introduced in the Section 3.2.2, the motion of electrons can be separated from that of nuclei. Thus the electronic eigenproblem described in Eq. (3.8) has to be solved for a fixed nuclear geometry. In that equation the assumed nuclear coordinates are treated as parame-

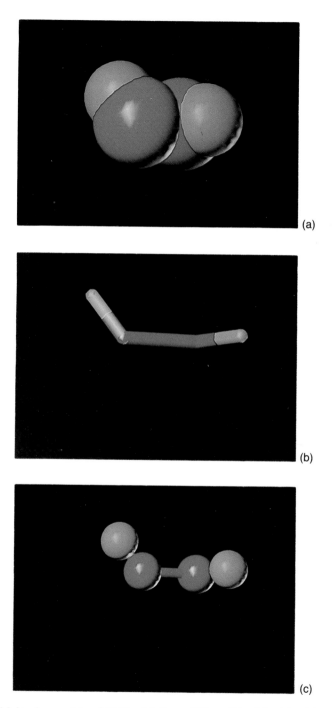

(a)

(b)

(c)

**Figure 1.3** Molecular models of $H_2O_2$: (a) Space-filling, (b) stick, (c) sticks-and-balls model.

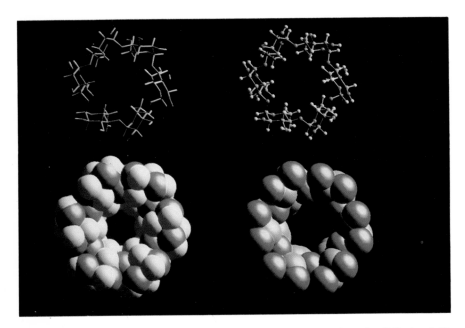

**Color Chart 1** Computer models of the most symmetric conformation of α-CyD: (top left) stick; (top right) stick-and-ball; (bottom) space-filling, showing all atoms (left) or only nonhydrogen ones (right).

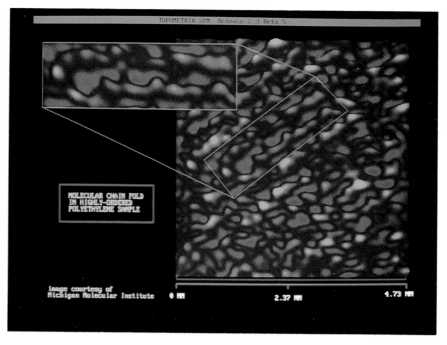

**Color Chart 2** AFM image of a highly ordered polyethylene sample showing molecular chain folding. Courtesy of Dr. S. Gehring of TopoMetrix GmbH.

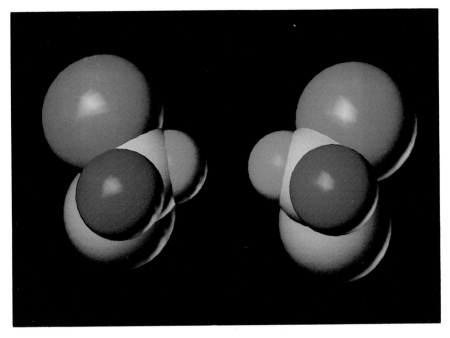

**Color Chart 3**  Computer models of two enantiomers of chloro-,bromo-,iodomethane.

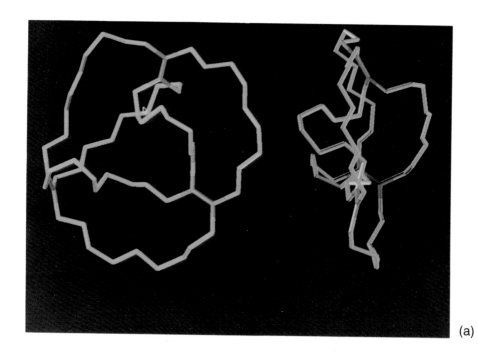

(a)

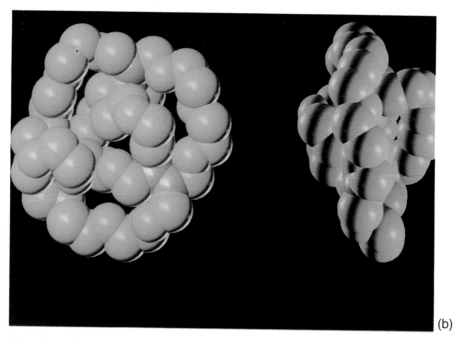

(b)

**Color Chart 4**   Two projections of computer models of a hypothetical hydrocarbon forming a knot. (a) Stick models and (b) space-filling models.

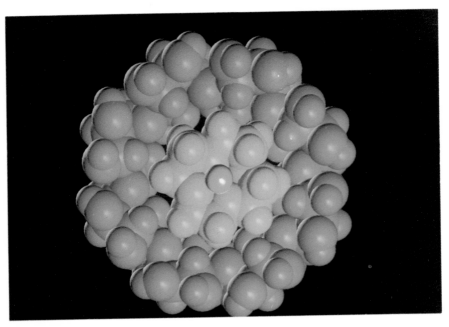

**Color Chart 5** A projection of a computer model showing a part of the complex formed by γ-CyD **14c,** crown ether **16,** and an alkaline ion.

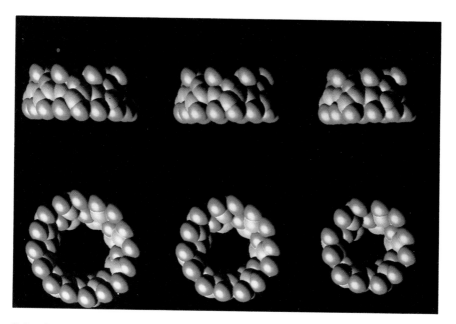

**Color Chart 6**   Top (bottom) and side (top) views of the most symmetrical structures of γ-, β-, and α-CyD exhibiting the diminishing size of the cavity.

ters. Solving Eq. (3.8) is still a formidable problem, since the electronic Hamiltonian operator $H_e$ depends not only on the electronic coordinates $r_i$ but also on the distances $r_{ij}$ between electrons $i$ and $j$. In accordance with the principles described in Section 3.2.1, the Hamiltonian for a hydrogen molecule is equal to

$$H = (\Delta_1^2 + \Delta_2^2) - (1/r_{a1} + 1/r_{a2} + 1/r_{b1} + 1/r_{b2} + 1/r_{12} + 1/R) \qquad (3.18)$$

where the indices 1,2 refer to electrons, $a,b$ denote atoms, $r_{a1}$ is the distance between electron 1 and nucleus $a$, etc., $r_{12}$ is the distance between the electrons 1 and 2, and the parameter $R$ is the internuclear distance. [The kinetic energy term of the nuclei, $(1/M)(\Delta_a^2 + \Delta_b^2)$ has been omitted in this Hamiltonian in accordance with the Born–Oppenheimer approximation. It is included in the nuclear Hamiltonian describing rotational and vibrational molecular motion.] In Eq. (3.18) for the hydrogen molecule, we have explicitly written the terms for electrons 1 and 2 and nuclei $a$ and $b$. For larger molecules they are replaced by summations over all nuclei and electrons.

According to the *valence-bond method* (VB), the wave function of the hydrogen molecule is a symmetric $\psi_s$ or antisymmetric $\psi_a$ combination of atomic hydrogen spin orbitals $1s$, $\phi_a$ and $\phi_b$:

$$\psi_s = 2^{-1/2} [\phi_a(1) + \phi_b(2)] \qquad (3.19)$$

$$\psi_a = 2^{-1/2} [\phi_a(1) - \phi_b(2)] \qquad (3.20)$$

The first of these functions is totally symmetric and has lower energy. When doubly occupied it represents the ground state of the hydrogen molecule, while the second, antisymmetric one describes the unoccupied, exited state. The symmetric function $\psi_s$ describing the bonding orbital has a minimum, while there is no minimum for the antisymmetric function describing the antibonding orbital. For molecules containing oxygen, nitrogen, and similar atoms, to a first approximation, the energy of nonbonding orbitals of lone electron pairs does not change upon bond formation. One of the definitions of the van der Waals radius of an atom associates it with the minimum of the symmetric wave function.

To satisfy the *Pauli principle* stating that only two electrons can occupy an orbital, atomic and molecular many-electron wave functions are often assumed in the form of the Slater determinant, that is, in the spin-orbital approximation

$$\Phi = (N!)^{-1/2} \begin{vmatrix} \psi_1(x_1) & \psi_2(x_1) & \dots & \psi_N(x_1) \\ \psi_1(x_2) & \psi_2(x_2) & \dots & \psi_N(x_2) \\ \dots & \dots & & \dots \\ \psi_1(x_N) & \psi_2(x_N) & \dots & \psi_N(x_N) \end{vmatrix} \qquad (3.21)$$

where $\psi_{2i-1} = \phi_i\alpha$ and $\psi_{2i} = \phi_i\beta$, $\phi_i$ is the spatial part of the orbital (identical in $\psi_{2i-1}$ and $\psi_{2i}$ for closed-shell systems) and $\alpha,\beta$ are spin functions corresponding to spin values $+\frac{1}{2}$ and $-\frac{1}{2}$. The assumption of a wave function in the form of a Slater determinant leads to the Hartree–Fock (HF) equation for the spin orbitals

$$f\psi_k = \epsilon_k\psi_k \qquad (3.22)$$

where $\epsilon_k$ can be interpreted as the $k$th orbital energy and $f(i)$ is an effective one-electron Fock operator assuming the form

$$f(i) = -\tfrac{1}{2}\Delta_i - \sum_{A=1}^{M} \frac{Z_A}{r_{iA}} + \vartheta^{\text{HF}}(i) \tag{3.23}$$

$\vartheta^{\text{HF}}(i)$ is the average potential experienced by the $i$th electron due to the presence of other electrons. Thus this term depends on the wave functions of other electrons.

### 3.2.4.1 Self-Consistent Field

The *self-consistent field* (SCF) method is used to solve Eq. (3.22). The initial form of $\psi$ spin orbitals is guessed first. Then the average potential is calculated, and the equation is solved for this potential. The new spin orbitals obtained by such a procedure are applied to obtain the new average potential, and the next approximation of spin orbitals and orbital energies is obtained. This approach is continued until self-consistency is achieved, that is, until HF potentials and spin orbitals in two consecutive steps do not differ. Thus a set of HF spin orbitals $\psi_i$ with orbital energies $\epsilon_i$ is obtained. For a system in the ground state containing $N$ electrons, the $N$ lowest-energy spin orbitals are occupied. The Slater determinant formed from these orbitals is the HF wave function. The unoccupied spin orbitals corresponding to higher energy are sometimes called *virtual spin orbitals*. In principle, the number of solutions of the HF equations is infinite. In practice, a finite set of spatial basis functions $\phi_\mu(q)$, $\mu = 1, 2, \ldots, K$, is used to solve it. Combining the spatial parts of the spin orbitals with the $\alpha$ and $\beta$ spin functions, one obtains Roothaan equations leading to $2K$ spin orbitals ($K$ with $\alpha$ spin and $K$ with $\beta$) and a complementary set of $(2K - N)$ unoccupied spin orbitals. The use of a larger and more complete set of basis functions $\phi_\mu$ leads to a greater degree of flexibility in the expansion of spin orbitals. According to the variational principle, it leads to a lower expectation value of the energy. The calculated energies are hence always larger than the lowest possible energy, called the HF limit. The acronyms HOMO and LUMO are often used for the highest occupied and lowest unoccupied molecular orbitals.

In the ab initio approach, all integrals in the formulae given are calculated explicitly. The expansion coefficients $c_{\nu i}$ for a molecular orbital in terms of atomic orbitals are obtained by solving the Hatree–Fock–Roothaan (HFR) equations mentioned

$$\sum_i c_{\nu i}(F_{\mu\nu} - \epsilon_i S_{\mu\nu}) = 0, \qquad i = 1, 2, \ldots, n \tag{3.24}$$

where $S_{\mu\nu}$ denotes the $\mu,\nu$ element of the overlap matrix equal to $\int \chi_\mu^* \chi_\nu \, dx$, $\epsilon_i$ is the energy of the $i$th orbital, and the $F_{\mu\nu}$ matrix elements are functions involving integrals including the $1/r_{12}$ operators.

Atomic spatial orbitals are usually assumed in the form of Gaussian-type orbitals

$$\varphi_{nl} = N_n \, x^k y^l z^m \, e^{-\alpha r^2} \tag{3.25}$$

As mentioned, the larger the basis set used, the more accurate the calculations. However, an increase can result in prohibitively long computer times, since the calculation time may increase as $N^4$, where $N$ is the size of the basis set. Thus, even with the rapid development of powerful computers, the size of the basis sets used poses a serious limitation.

The following notation for typical atomic basis sets was introduced in the GAUSSIAN program [18], which is in common use: (1) for hydrogen and carbon atoms minimal basis sets are limited to the use of one $1s$ orbital for each hydrogen atom and one $1s$, one $2s$, one $2p$ (meaning one $2p_x$, $2p_y$, $2p_z$) orbitals for each carbon atom. Extended bases may contain two or more $1s$, $2s$, and $2p$ functions with different exponents and, in addition, polarization functions, (of $p$ type for H, and of $d$ and $f$ type for second-row atoms). Contracted basis sets are now most frequently used, that is, the bases for which several Gaussian functions are combined with fixed coefficients into one. The molecular orbital expansion coefficients $c_{vi}$ are then calculated by solving the HFR equation in terms of contracted atomic orbitals [12]. Let us explain the contracted basis set notation using the example of the currently most-often-used 6-31G function. Here the number preceding the dash, 6, refers to the number of uncontracted Gaussian functions describing each core orbital. In the case of the carbon atom, six $s$-type functions represent the carbon inner-shell $1s$ orbital. The numbers after the dash denote orbitals in the valence shell. For carbon, "31" shows that there are two $s$ and $p$ orbital sets in the valence shell. In the first one denoted by "3," three $s$ ($p$) GTOs are contracted to describe a carbon valence $s$ and a valence $p$ orbital, respectively. In the second set denoted by the latter figure "1," an additional set is described by single uncontracted $s$ and $p$ Gaussians. For hydrogen "31" means that it has two $s$ orbitals. One of them consists of a linear combination of three $s$-type Gaussian functions, while the second is described by an uncontracted $s$ Gaussian.

Sometimes, time-consuming geometry optimization is carried out for a smaller basis set, such as 4-31G. Then the final energy and/or other values are calculated for a larger, 6-31G basis. Such a procedure is abbreviated as 6-31G/4-31G. The symbol 6-31G* denotes that, in addition to the functions used in the 6-31G basis set, $d$-polarization functions were used for carbon. A second star 6-31G** shows that $p$-polarization functions were used for hydrogen. In very accurate calculations, $d$ and $f$ atomic orbitals may be used for hydrogen, $f,g$ for carbon atom, etc. As said before, the larger the basis sets, the higher the accuracy and the higher the cost of the calculations.

### 3.2.4.2 Electron Correlation

The SCF approximation takes into account only the average interaction of the electrons in an atom or molecule. The corresponding Hamiltonian assumes the form

$$H_{SCF} = \sum_i (h_i + v_i) \tag{3.26}$$

where $h_i$ is the one-electron Hamiltonian and $v_i$ is the averaged potential of electron interactions. With a proper choice of basis set, the HFR method allows one to calculate the total molecular energy accurately to 0.3%. However, the total energy is very large, while comparing energies of two molecules or two molecular states we are often interested in small differences of these large values. In such cases, the correlation part of the Hamiltonian

$$H_{corr} = \sum_{i>j} \frac{1}{r_{ij}} - \sum_i v_i \qquad (3.27)$$

describing the instantaneous interaction between the electrons has to be taken into account. In Eq. (3.27) the $H_{corr}$ term is called the *fluctuation potential*.

The variational method used to estimate electron correlation consists of calculating the *configuration interaction* (CI) expansion. This can be carried out by taking into account not only the ground-state configuration described by one Slater determinant but also (excited) configurations characterized by other Slater determinants. Another approach consists of calculating the electron correlation using various levels of perturbation treatment. The second-order Moeller–Plesset procedure (MP2) is probably the most frequently used approximation in perturbational determinations of electron correlation effects.

Equation (3.23) can be approximately solved not only by ab initio but also by semiempirical methods. In the latter, specific values of certain integrals are assumed on the basis of spectral data, and some of them are set to zero. The methods, ranging from Extended Hückel (EHT) through CNDO and MINDO-type to AM1 are described in Ref. 19. Caution is advised in their use since they have often been parametrized to reproduce the results of specific experimental techniques for a given set of molecules.

In this chapter most attention is devoted to the LCAO method, which at present is most frequently used in QCs. It should be mentioned, however, that new, essentially different and very promising methods of density functionals are at present being developed [20].

## 3.2.5 The Applications of QCs and Their Significance

Virtually every molecular property can be computed by QCs. Standard applications with geometry optimization yield the energy and geometry of the molecule under investigation. Vibrational frequencies measured in IR spectra are also routinely calculated, to show that the established structure corresponds to the energy minimum [21], since imaginary values of the calculated frequencies denote the structures that are not minimum. Vibrational frequencies, force constants, and intensities of the IR bands are also calculated to solve specific problems [22]. The UV spectrum can also be easily obtained from QC calculations, since the bands correspond to transitions of an electron between the ground and an excited state. NMR chemical shifts and coupling constants [23], as well as many other molecular properties, can also be calculated fairly accurately. Standard program packages are now used in most theoretical studies in organic chemistry. As mentioned, the most frequently used for ab initio calculations is GAUSSIAN [18]. Semiempirical calculations are most frequently carried out using the MOPAC program package [24]. The programs enable not only calculations for known molecules but also allow one to take full advantage of studying hypothetical systems. The latter point can be illustrated by calculations for diagonally bonded cubane 65 [25] or faced-fused dicubane 66 [26] and some other unusual molecules to be discussed in Chapters 7 and 8. Today such calculations usually precede the syntheses of nonstandard molecules, allowing the proposal of new interesting systems as plausible synthetic targets. At present, the role of calculations as an independent valuable research tool is indisputable, but great problems still remain when the results of calculations are compared with the experimental ones.

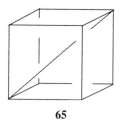

**65**

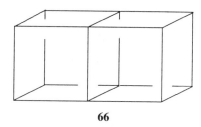

**66**

A comparison of the calculated and experimental results is generally a difficult task since (1) the latter are usually lacking. (2) If available, they often refer not to the molecule under study but to its derivative. (3) As discussed in Chapter 2, experimental results on spatial molecular structure are mostly obtained by X-ray analysis in the crystalline state, while the results of the calculations describe the isolated molecule. (4) As shown in Chapter 2, the results obtained using various experimental techniques differ even for bond lengths and angles. Even less accurate are experimental results for energy barriers. It should be stressed that the calculated values of the barriers reproduce the experimental ones, but the question of their origin is still not clear. On the basis of numerous calculations, one knows that it is a quantum, that is, nonclassical, effect. For instance, various quantum-chemical calculations at different levels of approximation have satisfactorily reproduced the barrier to internal rotation in ethane. However, they do not allow one to ascribe the hindering to a distinct interaction.

As said before, conformational energy differences and barriers are determined by QC as very small differences between very large numbers. Therefore, their accuracy is limited, and the method is usually more suitable for the study of constitutional isomers. For instance, QC for various isomers of $C_8H_8$ **7, 67,** and **68** are reviewed in Ref. 27.

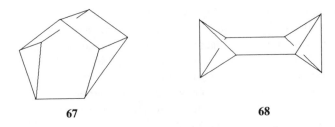

67                                              68

Another great problem concerning the applications of QC is that quantum theory treats a molecule as a whole. Therefore, as it is, its "language" is incompatible with that used in standard organic chemistry. Thus many useful chemical concepts such as the chemical bond, aromaticity, electronegativity, and π-stacking interactions, to name but a few, could not until recently be precisely and unambiguously expressed in terms of quantum chemistry. Work on this serious problem is in progress by Bader and co-workers [28]. It is illustrated by the discussion on the existence and strength of the central bond in [1.1.1]propellane **5** presented in Section 7.2.1 and the Bader treatment of the bond.

## 3.3 Molecular Mechanics

Contrary to QC, molecular mechanics (MM) [2, 29] does not have a sound theoretical foundation. Its only, but very impressive, justification is the amazing accuracy of the results obtained by using it. Within this method a hypothetical molecule without nonbonded interactions is assumed to have bond lengths $R_i^0$, bond angles $\vartheta_j^0$, and torsional angles $\phi_k^0$. In a real molecule in which nonbonded interactions operate, the corresponding values are $R_i$, $\vartheta_j$, and $\phi_k$. The steric energy $E_s$ of a real molecule is defined as the sum of increments associated with distortions of the bond lengths $E_R$, bond angles $E_\vartheta$, and torsional angles $E_\phi$, those describing nonbonded $E_{nb}$ and electrostatic $E_q$ interactions, and some other terms

$$E_s = \Sigma E_R + \Sigma E_\vartheta + \Sigma E_\phi + \Sigma E_{nb} + \Sigma E_q + \cdots \qquad (3.28)$$

where

$$E_R = \tfrac{1}{2} k_R (R_i^0 - R_i)^2$$

$$E_\vartheta = \tfrac{1}{2} k_\vartheta (\vartheta_j^0 - \vartheta_j)^2$$

and $k_b$ and $k_\vartheta$ are the corresponding force constants. The torsional increment $E_\phi$ is equal to $\Sigma V_n(1 - \cos n\phi)$, where $V_n$ is the barrier height of the $n$-fold barrier.

Lennard-Jones

$$E_{nb} = Ar^{-6} - Br^{-12} \qquad (3.29)$$

or Buckingham

$$E_{nb} = Ar^{-6} - B \exp(-Cr) \qquad (3.30)$$

van der Waals potentials described interactions of nonbonded atoms situated at a distance $r$. Large positive values of these terms for low $r$s correspond to strong interatomic repulsions. They do not allow two nonbonded fragments to come too close. At large distances a weak attractive dispersion interaction operates between the atoms. For an interaction of two atoms of the same type, the minimum between these extremes corresponds to the van der Waals radius of this type of atom.

The electrostatic term is expressed either as the Coulomb interaction of partial charges $q_i$, $q_j$ on atoms $i$ and $j$ or as the interaction of two dipoles. Cross terms of the types $(R_i^0 - R_i)(R_j^0 - R_j)$, $(\vartheta_i^0 - \vartheta_i)(\vartheta_j^0 - \vartheta_j)$, cubic terms $(R_i^0 - R_i)^3$, $(\vartheta_i^0 - \vartheta_i)^3$, the terms describing hydrogen bonding and some others are included in current force fields.

The choice of parameters in Eq. (3.28) is a difficult task, as discussed in detail in Refs. 2 and 29. The force constants $k_R$ and $k_\vartheta$ could in principle be taken from IR spectra; $V_n$ barriers, from MW studies; the $A,B,C$ constants in nonbonding interactions, from scattering experiments; $q_i$ charges, from quantum calculations. However, it should be emphasized that there is no precise way of obtaining $R_0$, $\vartheta_0$, and $\phi_0$ values. Moreover, it turned out that these parameters values taken from different experimental techniques do not yield satisfactory results. Therefore, another strategy is employed for their determination. A trial set of parameters is assumed on the basis of various considerations, and the steric energy $E_s$ is minimized as a function of molecular geometry. Then the calculated bond lengths and angles are compared with the corresponding experimental values for a trial set of molecules for which fairly accurate values have been determined. Next the assumed parameter values are changed to obtain better agreement between the calculated and experimental quantities. Such a procedure is continued until satisfactory agreement is reached.

Similarly, the steric energy differences between conformers, the calculated barriers, and the increments in additive schemes for the determination of heats of formation (HOF) can also be compared with and fitted to the experimental data. The important point is that the parameters determined in this way have no clear physical interpretation, and only the balance among them counts. For instance, two force fields, one having stronger H···H and weaker C···H interactions, and the second vice versa, would give very similar results. The minimization technique used in MM programs is known to allow only for a local minimum determination. Several procedures have been proposed to overcome this limitation, but no general solution to the problem exists.

A comparison of two different force fields was carried out in early work by Schleyer et al. [30]. The authors performed MM calculations for more than 120 saturated hydrocarbons, many of them hypothetical at that time. They have shown that for the prevailing majority of molecules under study, the calculated energies and geometries did not depend on the force field used. Few examples of the discrepancies referred to the strained-cage molecules. The sources of these differences are not clear even today. The largest of them was found for dodecahedrane **8**. It amounted to 45.5 kcal/mol and could not be checked against experiments in view of the lack of the latter value. As a matter of fact, to our knowledge an experimental value for **8** has still not been measured. The discrepancy between the experimental and calculated values is exceptional. As discussed in section 7.2.4.3, an unexpectedly very close reproduction of the experimental HOF value was obtained for cubane **7.**

One of the first examples of a successful joint application of MM calculations and ED was provided by the study of the cyclodecane **42** geometry mentioned in Chapter 2 [31]. The radial distribution curve obtained for this molecule could be satisfactorily reproduced by assuming this molecule to be present in only one conformation. Such an assumption seemed unjustified in view of its lability. Therefore, the MM calculations have been carried out for several conformations. Then the amount of each of them was determined, and the radial distribution curve calculated for the mixture. Interestingly, the latter curve did not differ significantly from that obtained based on the assumption of the presence of only one conformer. However, as mentioned, such an assumption is incompatible with the nonrigid structure of the ten-membered ring.

In view of the lack of sound theoretical foundations, MM was not considered credible by many researchers. However, the remarkable reproducibility of experimental results and its predictive power paved the way for a general acceptance of MM, which forms the basis of molecular modeling discussed in Chapter 11 [32]. Only a few examples of the application of the method will be mentioned here. The first one refers to the [3]rotane **69** geometry. The experimental X-ray study by Prange et al. [33] yielded the largest length for the *b* bond, and this result could not be reproduced by theoretical studies, in particular by the MM calculations [34]. A redetermination of the experimental data [35] was undertaken, which yielded excellent agreement with the MM calculated values presented in Table 3.1.

A very revealing example of the successful application of MM calculations is provided in the work of Biali et al. [36]. The authors synthesized decakis(dichloromethyl)biphenyl **70,** determined its structure, separated enantiomers, and studied its

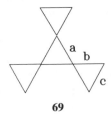

**69**

**Table 3.1**   The Experimental and Calculated Values of the Bond Lengths
            in [3]rotane **69** in pm

| Method\bond | *a* | *b* | *c* | Ref. |
|---|---|---|---|---|
| X-ray | 147.5(4) | 150.4(4) | 149.6(4) | 33 |
| MM | 147.5 | 147.7 | 153.8 | 34 |
| X-ray | 147.5(2) | 147.7(1) | 152.5(2) | 35 |

racemization. The X-ray structure of its 1:1 adduct with tetrahydrofurane revealed some peculiar features. In particular, its one aromatic ring was planar, while the second one assumed a twist-boat conformation, and there was an angle of 165.2° between the *para* positions and the midpoint of the central bond. The MM calculations on the isolated molecule have proved that these unusual features are an intrinsic property of **70** and are not caused by the crystal forces.

$$Cl_2HC \quad Cl_2HC \quad CHCl_2 \quad CHCl_2$$

$$Cl_2HC \quad CHCl_2$$

$$Cl_2HC \quad Cl_2HC \quad CHCl_2 \quad CHCl_2$$

**70**

Because of their conceptual simplicity and reliability, MM calculations have become a standard technique, not only for theoreticians, but also for organic synthetic chemists. As discussed in Chapter 11, they form a part of all molecular modeling packages. As shown in Ref. 37, MM allows not only for rationalization of experimental trends, but also for the prediction of properties of new hypothetical molecules. For instance, octahedrane **71** was predicted to be one of the most stable members of the $C_{12}H_{12}$ family of saturated cage compounds on the basis of MM calculations [38]. Thus it was proposed as a plausible synthetic target that should exhibit the unusually high value of 74° for the torsional CCCC angle in the central cyclohexane ring. A few years later **71** was indeed synthesized [39], and its X-ray analysis [40] showed remarkable agreement with the value calculated using the MM2 force field [38] developed for standard structures. Some other examples of the MM applications are presented in Chapters 2, 7, 8, and 10.

**71**

Future developments in the field will undoubtedly include the introduction and refinement of new force fields for specific applications (for sugars, peptides, etc.) and the incorporation of new atoms to extend their applicability. However, a proper description of mobile supramolecular systems such as cyclodextrin **14** complexes, peptides, etc. calls for applications of dynamic simulations methods.

## 3.4 Dynamics Simulations

Dynamics simulation (DS) methods are a natural extension of MM. They are especially useful in studies of flexible systems exhibiting numerous low-lying conformations separated by low barriers. Such a situation is typical of biomolecules and polymers, as well as of cyclic systems with a coupling of degrees of freedom. Two approaches are adopted in investigations of these systems, namely, molecular dynamics [3a] and Monte Carlo simulations [3b].

### 3.4.1 Molecular Dynamics

In molecular dynamic (MD) simulations, Newton's equations of motion are used:

$$\frac{d^2 \bar{x}_i}{dt^2} = \bar{a}_i = \frac{\bar{F}_i}{m_i} \tag{3.31}$$

$$\bar{F}_i = -\text{grad}_i(V) \tag{3.32}$$

where $\bar{x}_i(t)$ is the trajectory describing the time evolution of the system under study, $\bar{a}_i$ denotes acceleration, $\bar{F}_i$ is the force, $\text{grad}_i$ is the gradient, and $V$ is the potential function of the system.

In MD simulations the potential functions are assumed in the simplified form of those used in MM calculations described in the former section with no cross terms used. In addition, separate terms taking into account the hydrogen bonding energy are often included in the studies of nucleic acids, proteins, and peptides.

Newton's equations (3.31) and (3.32) are solved by discrete integration over a large number of finite time steps ($dt \rightarrow \Delta t$). According to statistical mechanics, the kinetic energy of the system defines its temperature $T$

$$\tfrac{3}{2} NkT = \tfrac{1}{2} \sum_{i}^{N} m_i \bar{v}_i^2 \tag{3.33}$$

where $N$ is the Avogadro numer, $k$ the Boltzmann constant, and $m_i$ the atomic masses.

The time evolution can be also expressed in terms of the velocities of the atoms $\bar{v}_i(t)$. In this case, the model temperature is kept constant by the velocities' scaling, corresponding to coupling the system to a heat bath. The simulations can also be carried out under conditions of constant pressure.

MD simulations can be performed not only for individual molecules, but also for molecular ensembles consisting of a molecule or host–guest complex with solvent. In the latter case, the molecules are arranged in a cell of appropriate volume (for liquids it is usually a cube; for crystals a unit cell) to obtain a desired density. Periodic boundary conditions are then imposed to eliminate surface artefacts and generate a quasi-infinite number of particles. For systems with many internal degrees of freedom, the optimum step $\Delta t$ for the simulation is $10^{-15}$–$10^{-14}$ seconds. With supercomputers it is possible to carry out the simulations over $10^6$ time steps, enabling a reliable study of internal rotations and aggregations of systems containing ca. 100 atoms [3a].

Rapid, periodic movements and slow nonperiodic ones can be analyzed by considering the trajectory of the system. Average structures and energies as well as correlations among various types of movement can be elucidated from it. There seem to be three restrictions to such studies. The first one is associated with the accuracy of the force field used. Contrary to the situation with MM force fields, for which comparative studies on the same sets of molecules exist (e.g., Ref. 30), the operation of GROMOS [41a], AMBER [41c], SYBYL [41b] or QUANTA/CHARMm [41d] MD force fields seem not to have been extensively checked against each other. The second restriction refers to the protocol of calculations since the cited programs differ in the computational procedures used. For instance, contrary to MM programs, those for MD simulations allow for an interactive choice of minimization procedures, of the methods used to calculate partial charges, etc. This unables the use of MD programs as black boxes and requires a much deeper knowledge of the methods. The last limitation comes from the limited computer power available today, limiting the size and number of time steps used. van Gunsteren and Berendsen [3a] claimed in 1990 that reliable dynamic simulations can only be carried out for systems consisting of fewer than 100 atoms.

As discussed in Section 2.2.2, dynamic simulations for $\alpha$-cyclodextrin **14a** were carried out by van Gunsteren et al. [42] in aqueous solutions and the crystalline state. We believe that for molecules of this size reliable MD calculations require considerably larger simulation times than the standard 100 ps usually used.

## 3.4.2 Monte Carlo Method

Monte Carlo simulations [3b] are based on the calculation of ensemble averages of the properties of fluctuating systems. Random numbers are used to generate $n$ local variations of the system, such as the displacement of an atom, the change in torsional angle, etc., producing its new state. If the energy of a new state is lower than that of the previous one, the next variation will start from this new state. Otherwise, for higher energy, the Boltzmann factor $f$ equal to $\exp(E^{new} - E^{old})/kT$ is calculated. Then the $f$ value is compared with another random number. If the random number is smaller than $f$, the previous state is retained. If not, the new state is adopted. $f$ depends only on the directly preceding state and plays the role of the transition probability for a state of higher energy. Such fluctuations corresponding to a random walk through conformational space are called a *Markov chain* [43]. They allow the system to fluctuate in accordance with a Boltzmann distribution for temperature $T$. Similarly to MD, the Monte Carlo method is mainly used in the study of liquids and biopolymers.

## 3.5 Dry Chemistry versus Wet Chemistry. Proposing Novel Interesting Molecules as Plausible Synthetic Targets

To summarize, due to the rapid development of computers and computer programs, theoretical calculations have developed into an important tool in chemical research. In contrast to syntheses constituting "wet chemistry," they are sometimes called "dry chemistry." Using highly efficient computers and user-friendly programs, it is sometimes possible to obtain results that their ease of being obtained and accuracy rival the corresponding experimental data. On the other hand, the attractiveness of the graphical presentation of organic molecules and biomolecules and the ease of using the program sometimes results in a lack of understanding of the theory underlying the calculations and their limitations. When properly applied, the calculations provide models enabling an understanding of existing data and prediction of novel interesting results.

One of the most spectacular aims of theoretical studies is proposing new prospective synthetic targets. As discussed in Sections 3.2.5 and 7.2.1, the stability of apparently highly strained [1.1.1]propellane 5 and its properties have been predicted by Wiberg et al. on the basis of QCs [9]. The synthesis carried out later by the same group proved the prognostications. As discussed in Section 3.3, the feasibility of octahedrane 71 and its geometry were predicted prior to its synthesis. As will be discussed in Section 7.2.2, even when the calculations yield conflicting results, as was the case with the energy of planarization of methane, they inspired the syntheses of numerous fascinating fenestranes 63 [44]. On the basis of QCs [45], we know that the central atom in windowpane cannot adopt a planar configuration. However, we believe that today's massive theoretical calculations will lead to the synthesis of a molecule having a pyramidal carbon atom in the near future.

### References

1a. P. W. Atkins, *Molecular Quantum Mechanics. An Introduction to Quantum Chemistry*, Clarendon Press, Oxford, 1970.

1b. A. Szabo and N. S. Ostlund, *Modern Quantum Chemistry. Introduction to Advanced Electronic Structure Theory*, McGraw-Hill, New York, 1989.

2. E. Osawa and H. Musso, *Angew. Chem., Int. Ed. Engl.*, **22** (1983) 1; N. L. Allinger, *Adv. Phys. Org. Chem.* **13** (1976) 1.

3a. W. F. van Gunsteren and H. J. C. Berendsen, *Angew. Chem. Int. Ed. Engl.*, **29** (1990) 992.

3b. K. Binder, Ed., *Monte Carlo Methods in Statistical Physics*, Springer, Berlin, 1979; B. J. Berne, Ed., *Statistical Mechanics*, Plenum Press, New York, 1977.

4. As visualized by the MM versus FF controversy, as well as by MD versus DS, the terminology in the field is not fully settled yet. Surprisingly, issue No. 7 of *Chemical Reviews* **93** (1993), bearing the term *molecular mechanics* in the title, contains an article on molecular dynamics. We believe that such equivocality should be avoided.

5. E. Osawa, Kagaku (Kyoto) **25** (1970) 854 (in Japanese); *Chem Abstr.*, **74** (1971) 75698w; Z. Yoshida and E. Osawa, *Aromaticity*, Kagakudoin, Kyoto, 1971 (in Japanese).

6.  D. A. Boc'var and E. G. Gal'pern, *Dokl. Akad. Nauk SSSR* **209** (1973) 610 [English transl. *Proc. Acad. Sci. USSR* **209** (1973) 239].

7a. R. Hoffman, R. W. Alder, and C. F. Wilcox, *J. Am. Chem. Soc.* **92** (1970) 4992; R. Hoffmann, *Pure Appl. Chem.* **28** (1971) 181.

7b. S. Durmaz, J. N. Murrell, and J. B. Pedlay, *J. Chem. Soc., Chem. Commun.* (1972) 933; J. N. Murrell, J. B. Pedlay, and S. Durmaz, *J. Chem. Soc., Faraday Trans. 2* **69** (1973) 1370.

7c. D. C. Crans and J. P. Snyder, *J. Am. Chem. Soc.* **102** (1980) 7153.

8.  B. R. Venepalli and W. C. Agosta, *Chem. Rev.* **87** (1987) 399.

9.  K. B. Wiberg, *Acc. Chem. Res.* **17** (1984) 379.

10. S. Wilson, *Chemistry by Computer,* Plenum Press, New York, 1986.

11. T. Clark, *A Handbook of Computational Chemistry,* J. Wiley, New York, 1985, p. 65.

12. Ref. 1b, p. 41.

13. In particular, nuclear spin is not related to the rotating magnetic moment of the nucleus. If it were, the neutron would not have any spin. In reality, it has no magnetic moment but does have a spin of $\frac{1}{2}$.

14. V. Dashevskiy, *Conformational Analysis of Organic Molecules* (in Russian), Khimya, Moscow, 1982, p. 19.

15. K. S. Pitzer, *Chem. Rev.* **27** (1940) 39.

16. P. Kollmann, *Chem. Rev.* **93** (1993) 2395.

17. E. L. Eliel, N. L. Allinger, S. J. Angyal, and G. A. Morrison, *Conformational Analysis,* Interscience Publishers, New York, 1965, Section 2-3.B.

18. GAUSSIAN, Inc., Carnegie Office Park, Building 6, Pittsburgh, PA 15106.

19. G. A. Segal, Ed., *Semiempirical Methods of Electronic Structure Calculation,* Parts A and B, Plenum, New York, 1977.

20. R. M. Dreizler and J. da Providencia, Ed., *Density Functional Methods in Physics,* Plenum Press, New York, 1985.

21. For a system composed of *N* atoms, there are 6 zero frequencies corresponding to the translational and rotational movement of the whole system. If the optimized geometry corresponds to an energy minimum (not necessarily the absolute one), then all calculated frequencies are greater than zero. Otherwise, it corresponds to a maximum when all frequencies assume negative values or to a saddle point for which one of the frequencies is less than zero.

22. See, for instance, K. B. Wiberg, S. T. Waddell, and R. E. Rosenberg, *J. Am. Chem. Soc.* **112** (1990) 2184.

23. M. Bremer, K. Schotz, P. R. von Schleyer, V. Fleischer, M. Schindler, W. Kutzelnigg, W. Koch, and P. Pulay, *Angew. Chem., Int. Ed. Engl.* **28** (1989) 1042; P. R. von Schleyer, J. W. de Carneiro, W. Koch, and K. Raghavachari, *J. Am. Chem. Soc.* **111** (1989) 5475; V. Galasso, *Chem. Phys.* **181** (1994) 363.

24a. MOPAC is a public domain semiempirical program package written by James J. P. Stewart at Frank J. Sailer Research Laboratory, United States Air Force Academy, Colorado Springs, CO 80840. It can be used to carry out calculations using MNDO, MINDO/3, AM1, and PM3 methods.

24b. MOPAC is also distributed by Quantum Chemistry Program Exchange, Creative Arts Building 181, Indiana University, 840 State Highway 46 Bypass, Bloomington, IN 47405. QCPE also provides more than 600 other programs for QC, MM, and some other calculations.

25.  K. Hassenrück, J. G. Radziszewski, V. Balaji, G. S. Murthy, A. J. McKinley, D. E. David, V. M. Lynch, H.-D. Martin, and J. Michl, *J. Am. Chem. Soc.* **112** (1990) 873; D. A. Hrovat and W. T. Borden, *ibid.* **112** (1990) 875; P. E. Eaton and J. Tsanaktsidis, *ibid.* **112** (1990) 876.

26.  E. T. Seidl and H. F. Schaefer, Jr., *J. Am. Chem. Soc.* **113** (1991) 1915.

27.  K. Hassenrück, H.-D. Martin and J. R. Walsh, *Chem. Rev.* **89** (1989) 1125.

28.  R. F. W. Bader, P. L. A. Popelier, and T. A. Keith, *Angew. Chem., Int. Ed. Engl.* **33** (1994) 620.

29.  J. E. Williams, P. J. Stang, and P. R. von Schleyer, *Ann. Rev. Phys. Chem.* **19** (1968) 531.

30.  E. M. Engler, J. D. Andose, and P. von R. Schleyer, *J. Am. Chem. Soc.* **95** (1973) 8005.

31.  Ref. 25 from Chapter 2.

32.  H. Frübeis, R. Klein, and H. Wallmeier, *Angew. Chem., Int. Ed. Engl.* **26** (1987) 403.

33.  T. Prange, C. Pascard, A. de Meijere, U. Behrens, J.-P. Barnier, and J. M. Conia, *Nouv. J. Chim.* **4** (1980) 321; *J. Chem. Soc., Chem. Commun.* **9** (1979) 425.

34.  H. Dodziuk, unpublished results, 1982; A. J. Yoffe, V. A. Svyatkin, and O. M. Nefedov, *Izv. AN SSSR, Ser. Khim.* **4** (1987) 801.

35.  R. Biese, T. Miebach, and A. de Meijere, *J. Am. Chem. Soc.* **113** (1991) 1743.

36.  S. E. Biali, B. Kahr, Y. Okamoto, R. Aburatani, and K. Mislow, *J. Am. Chem. Soc.* **110** (1988) 1917.

37.  H. Dodziuk, *Top. Stereochem.* **21** (1994) 351.

38.  H. Dodziuk and K. Nowinski, *Bull. Pol. Acad. Sci., Chem.* **35** (1987) 195; H. Dodziuk, *ibid.* **38** (1990) 11.

39.  C. H. Lee, S. Liang, T. Haumann, R. Boese, and A. de Meijere, *Angew. Chem., Int. Ed. Engl.* **32** (1993) 559.

40.  A. de Meijere, private communication.

41a.  GROMOS distributed by Biomos B. V., Laboratory of Physical Chemistry, University of Groningen, Nijenborgh 16, 9747 AG Groningen, The Netherlands.

41b.  SYBYL distributed by Tripos Associates, 1699 Hanley Road, Suite 303, St. Louis, MO 63144-2913.

41c.  AMBER distributed by Dr. P. A. Kollman, Department of Pharmaceutical Chemistry, University of California, San Francisco, CA 94143.

41d.  QUANTA/CHARMm distributed by Molecular Simulations, Inc., 16 New England Executive Park, Burlington, MA 01803-5297.

42.  G. G. Lowry, Ed., *Markov Chains and Monte Carlo Calculations in Polymer Science*, Dekker, New York, 1970.

43.  J. Köhler, W. Saenger, and W. F. van Gunsteren, *J. Mol. Biol.* **203** (1988) 241.

44.  B. R. Venepalli and W. C. Agosta, *Chem. Rev.* **87** (1087) 399.

45.  J. M. Schulman, M. L. Sabio, and R. L. Disch, *J. Am. Chem. Soc.* **105** (1983) 743.

# CHAPTER
# 4

# Molecular Symmetry

As in many other places in this book, a molecule is treated here as an ideal set of point masses connected by bonds. This is, obviously, only a simplified model that is very useful, even indispensable, in most cases. However, it can be an oversimplification for real molecules moving as a whole and undergoing internal motion. The very small but nonvanishing dipole moment of methane (discussed in Chapter 2) exemplifies this point, since it is incompatible with its $T_d$ symmetry and the equivalence of all four CH bonds. The very small, but detectable, dipole moment results from molecular distortions that lower its symmetry. It clearly shows the difference between idealized and real structures.

Such differences between mathematical models of a chiral molecule and the molecule itself are discussed in detail by Mislow and Bickart [1]. In many respects, their reasoning is applicable to the problem of symmetry. Many excellent books on the mathematical aspects of molecular symmetry and on the implications it has on the physicochemical properties of a molecule under study [2, 3] have been published. Here a practical, introductory approach will be adopted with a special emphasis on the difference between the symmetry of a conformation and the average molecular symmetry.

Local symmetry has to be distinguished from the overall one, which, in turn, can be different from the symmetry pertaining to rotational barriers. Let us examine Fig. 4.1 to visualize the differences in toluene **62** ($X = H$). Both (a) and (b) conformations presented have a symmetry plane, but in the former one hydrogen atoms in the 2 and 6 positions (and those in the 3 and 5 positions) are equivalent, while in the latter the hydrogen atoms in the above positions are different. The local symmetry of the methyl group in toluene with a threefold symmetry axis is different from the

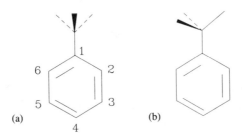

**Figure 4.1**  Different conformations of the methyl group in toluene. (a) The symmetry plane is perpendicular to the aromatic ring plane and goes through the C1 and C4 atoms. (b) The aromatic ring lies in the symmetry plane.

overall one. However, the internal rotation of the methyl group in toluene has a (very low [4]) sixfold barrier brought about by the simultaneous existence of the local threefold axis and the plane of symmetry.

Only point symmetry groups will be discussed here, that is, the groups, symmetry elements of which leave at least one point unchanged. Schönflies notation [5a] will be used to describe molecular symmetry. Crystallographic symmetry groups [5b] will not be covered here since (1) they involve translational symmetry in addition to the point group symmetry of an individual molecule and (2) they describe infinite sets of molecules.

Symmetry is one of the most important properties of a molecule. Due to the symmetry, the number of independent geometrical parameters describing the structure of a molecule under study diminishes, and the corresponding experimental data (such as IR, MW, NMR spectra as well as X-ray diffractograms, etc.) are considerably simplified. Let us consider fullerene $C_{60}$ **11**. The molecule has icosahedral symmetry $I_h$ (see below), and all its carbon atoms are equivalent. Therefore, as discussed in Chapter 2, the $^{13}C$ NMR spectrum of this molecule exhibits only one signal, which proved the fullerene structure at an early stage of its study [6a]. Due to its high symmetry, the equivalent carbon atoms in $C_{60}$ are connected solely by two types of bonds. Accordingly, as described in Chapter 2, using NMR, only two CC bond lengths have been found for **11** [6b]. The mutual exclusion principle discussed in Section 2.3.4 illustrates a simplification of vibrational spectra due to the presence of an inversion center. These examples show that molecular symmetry has serious practical consequences, allowing for the speed and precision of analysis of experimental results and theoretical calculations.

As mentioned earlier, first we will treat a molecule as a static system of point masses. The molecule is said to belong to a certain symmetry group if the execution of an operation of this group yields a molecule equivalent to the starting one. The following symmetry elements are used in the Schönflies system: the $n$-fold rotational axes, $C_n$; the center of symmetry, $i$; symmetry planes, $\sigma$; improper axes, $S_n$; and identity, $E$. The last element corresponds to the situation in which the application of a symmetry operation yields as the result a system indistinguishable from the starting one. It is added to ensure that symmetry operations form a group in the sense of group theory. This means that a sequence of symmetry operations must also be an operation belonging to the same group (see below).

A molecule has a symmetry axis $C_n$ if the rotation around this axis by the angle $360°/n$ [or by the angles $360° \times p/n$, where $p$ is equal to $1, 2, \ldots, (n-1), n$] yields a molecule equivalent to the starting one. For instance, in a molecule having a $C_3$ symmetry axis (coinciding with the CH bond in chloroform $CHCl_3$ in Fig. 4.2), rotations around the axis by the angles $120°$, $240°$, and $360°$ produce [Fig. 4.2(a)] the same molecule with different atomic numbering. The last value describes rotation by $360° \times 3/3$ or $0°$, which does not produce any rotation and corresponds to the identity operation $E$. Similarly, reflecting a molecule possessing a symmetry plane $\sigma$ [such as HOD, shown in Fig. 4.2(b)] in this plane produces an equivalent structure. Inversion of all atoms in the center of symmetry $i$ [situated at the midpoint of the C=C bond in ethylene, shown in Fig. 4.2(c)] can easily be carried out if the molecule is put into a coordinate system with the origin at the center of symmetry. Then changing signs of all Cartesian coordinates of the atoms corresponds to inversion. Improper axes $S_n$ are present in the molecule if rotation by an angle $360°/n$ with a subsequent reflection in a plane perpendicular to the axis yields a molecule indistinguishable from the starting one. It can be shown that for odd values of $n$, the $S_n$ axis is equivalent to the simultaneous presence of a $C_n$ axis and a symmetry plane $\sigma$. For even $n$ values the $S_n$ axis is equivalent to the presence of a $C_{n/2}$ axis and a center of symmetry. Thus $S_1$ is equivalent to a symmetry plane $\sigma$, and $S_2$ is equivalent to a center of symmetry $i$. Cyclohexane in a chair conformation **22b** can serve as an example of a molecule having an $S_6$ symmetry axis. This axis goes through the molecular center parallel to the axial CH bonds.

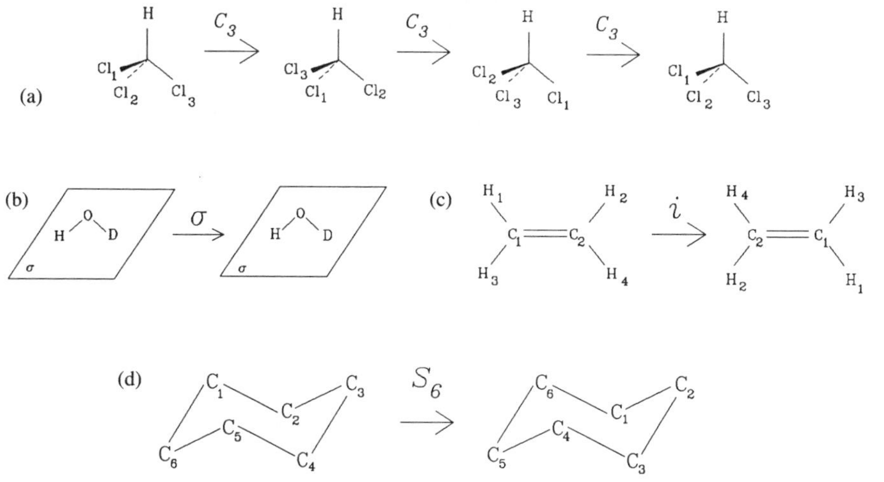

**Figure 4.2** Examples of symmetry operations. (a) Threefold axis in $CHCl_3$. (b) The symmetry plane in HOD. (c) The inversion center in $H_2C{=}CH_2$. (d) The $S_6$ axis in cyclohexane.

A molecule can have more than one element of symmetry. The elements form a group in the mathematical sense if (1) a product of two elements of the group, that is, a consecutive application of these elements, is also an element of the group. Let us note that the order of application of the symmetry operations is usually of importance. (2) The product of the elements in a group are always associative. This means that for a given order of application, symmetry operations can be grouped in a different way, and this does not change the final result of their application. Let us look at ethylene, presented in Fig. 4.2(c). The molecule has, among other symmetry elements, two symmetry planes: $\sigma_1$, perpendicular to the double bond, and $\sigma_2$, in which the bond lies. It also has an inversion center $i$. As can be easily checked, the subsequent execution of three operations, $\sigma_1$, $\sigma_2$, and $i$, can be carried out in two different ways

$$(\sigma_1\sigma_2)i = \sigma_1(\sigma_2 i)$$

which both yield the identity operation $E$. The latter one does not introduce any changes in the system under investigation. (3) For each element of a group $A$ there is always an inverse or reciprocal element $A^{-1}$ such that the consecutive application of the element and its inverse yields the identity operation $E$:

$$A \times A^{-1} = A^{-1} \times A = E$$

In the following the point symmetry groups will be listed with examples of molecules possessing them.

Symmetry groups $C_n$: If a molecule has only a rotational axis $C_n$ of order $n$ and identity element $E$, then it belongs to the symmetry group $C_n$. For the sake of completeness, the symbol $C_1$ denotes a symmetry group, thus the symmetry of a molecule lacking other than identity symmetry elements. Methane with four different substituents (such as CHClBrJ **27**), nonplanar **72** [7], monodeuterated hydrogen peroxide **73a**, 1,2,3-tri-t-butylnaphthalene **74** with a nonplanar carbon skeleton [8], and two isomers of the quaternary ammonium salts **75c** and **75d** synthesized by McCasland and Proskow [9], which will be discussed in the next chapter, are examples of molecules that do not have a symmetry element. They all belong to the group $C_1$. It should be stressed that most molecules have no symmetry and belong to this group.

$C_1$

72            73a X=D            74

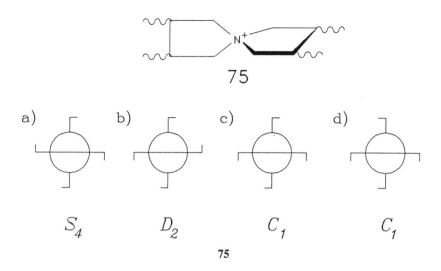

75

a) $S_4$  b) $D_2$  c) $C_1$  d) $C_1$

75

Hydrogen peroxide **73b** itself [10], hexahelicene **76** [11], and *trans*-cyclooctene **77** [12] have only one $C_2$ axis and belong to the group **C₂**.

$C_2$

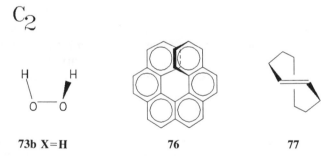

**73b X=H**      **76**      **77**

**78** [13] in a conformation depicted in the formula as well as the rigid molecules **79** [14], and **80** [15], have a $C_3$ symmetry axis as a sole symmetry element and belong to the **C₃** group. **81** has the same symmetry if the macrocycle assumes a nonplanar symmetric conformation [16]. Its average symmetry is higher (**C₃ₕ**). A more complicated situation is encountered by tri-O-thymotide **82** [17]. In solution the molecule is nonrigid, and it adopts several conformations that are in dynamic equilibrium. As a result, the average structure of **82** belongs to the **C₃** symmetry group. As will be discussed in detail in the next chapter, molecules with such a symmetry are chiral, and two enantiomeric structures are possible for them. Actually, **82** was the first molecule from this symmetry group to be resolved into enantiomers [18].

$C_3$

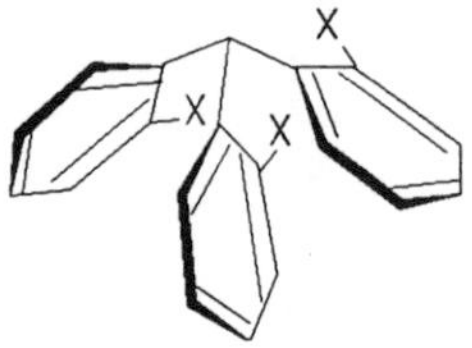

**78** 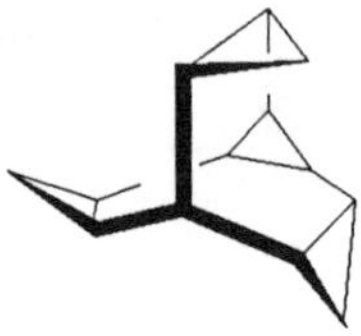      **79** 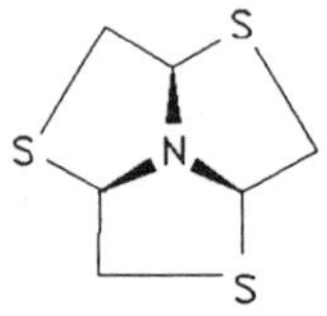      **80**

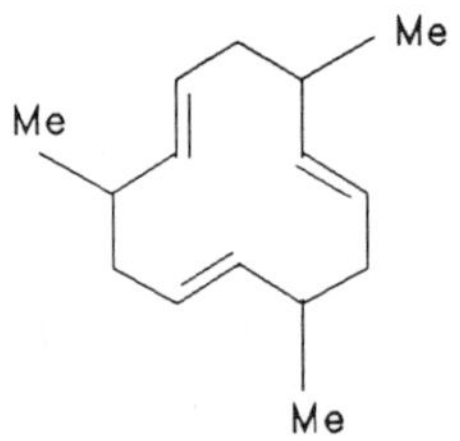

**81** 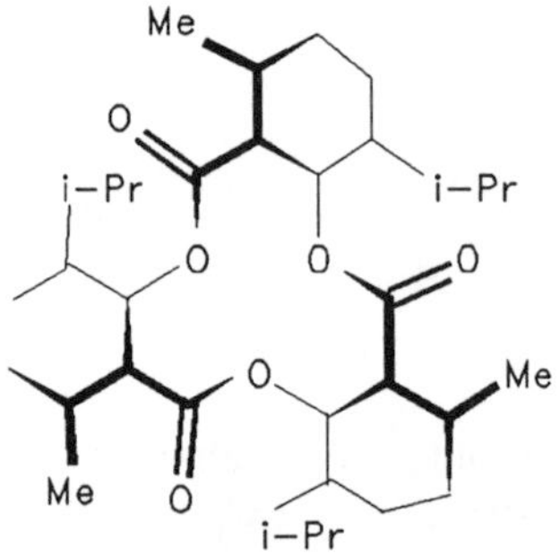      **82**

Organic molecules belonging to $C_n$ groups with higher $n$ values are very rare. For instance, we were not able to find any having $C_4$ symmetry. Averaged structures of α-, β-, and γ-cyclodextrins **14a–c,** respectively, belong to $C_6$, $C_7$, and $C_8$ symmetry groups, respectively. As discussed in detail in Section 10.3, the fact that the symmetry of cyclodextrins refers to the averaged structures is usually overlooked.

Symmetry groups $C_s$ and $C_i$: To the first belong molecules that have a symmetry plane as a sole symmetry element. Nonlinear ONCl **83** [19] and $CF_3C{\equiv}CSF_3$ **84** [20], HOD and a dimer **85** [21] of enantiomers of lactic acid belong to the $C_s$ group. Similarly, **86** in the *trans* conformation and molecules **87** and **88** [21] belonging to the second group have only a center of symmetry *i*.

$C_S$

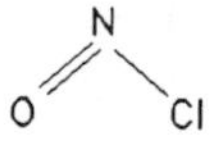

**83** 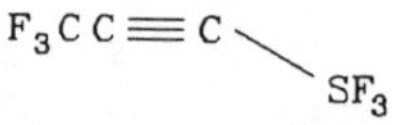      **84** 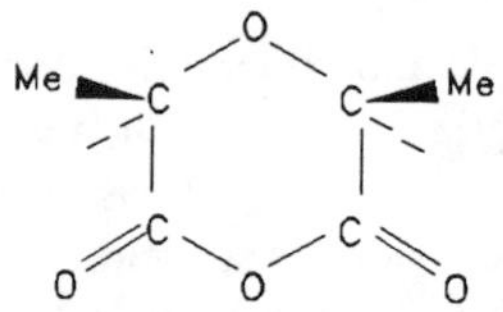      **85**

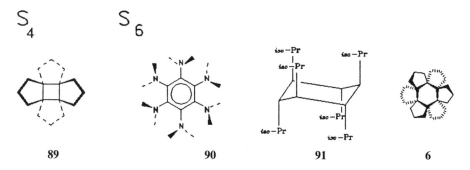

**86**  **87**  **88**

Symmetry groups $S_n$: Molecules belonging to these groups have only an improper $S_n$ axis. It should be noted that in this case $n$ can assume only even values, since for odd ones the existence of such an axis in a molecule corresponds to the simultaneous presence of a $C_n$ axis and the symmetry plane perpendicular to this axis. As mentioned earlier, the group $S_1$ contains a symmetry plane only; thus it is equivalent to the group $C_s$. Similarly, the group $S_2$ contains only a center of symmetry $i$ and is equivalent to the group $C_i$. The symmetry group $S_4$ (symmetry elements: $S_4$ axis and $C_2$ axis overlapping with the former one) has one isomer of the quaternary ammonium salts **75a** [9]. As will be discussed in the next chapter, due to the presence of an improper $S_4$ axis, this molecule is not chiral, contrary to three other isomers **75b–d,** which do not have improper axes and, consequently, are chiral. The synthesis of all four isomers **75** and assignment of their symmetry and chirality played an essential role in the formulation of the symmetry condition of molecular chirality. Cyclobutane **47,** with a nonplanar conformation of the four-membered ring discussed in Chapter 2, one conformation of cyclooctatetraene **21,** which undergoes rapid ring inversion [22], and approximately substituted derivatives of **89** [23] also have $S_4$ symmetry. The $S_6$ symmetry group contains [6.5]coronane **6** [24], which has a chair conformation of the central cyclohexane ring. The same symmetry is exhibited by hexaamine **90** [25], with amine groups regularly twisted by 13⁰, and all-*trans*-1,2,3,4,5,6-hexaisopropylcyclohexane **91** [26], which, somewhat unexpectedly, has all its substituents in axial positions.

**89**  **90**  **91**  **6**

The symmetry groups $C_{nv}$: If a molecule has a symmetry axis $C_n$ and $n$ symmetry planes $\sigma_v$ ("v" stands for vertical) intersecting at the $C_n$ axis, then this molecule belongs to the $C_{nv}$ group. For obvious reasons, the group $C_{1v}$ is identical with the group $C_s$ discussed earlier. $H_2O$, $CH_2Cl_2$, and the molecules **92–95** [27–30] belong to the $C_{2v}$ symmetry group. **93–95** are of interest as prospective new materials to be used in electronics (see Section 12.2 discussing such applications). Pyramidal ammonia $NH_3$, $CH_3Cl$, molecules **96** [31] and **97** [32], and monosubstituted adamantanes **98** ($X = Y = Z =$ H) [33] have $C_{3v}$ symmetry. Interestingly, molecules with $C_{4v}$ symmetry are very rare. We were able to find only **99** [34], **100** [35], and the average structure of the most symmetrical conformation of [4]calixarene **15** [36] with this symmetry, since cyclobutane **47** (discussed in Chapter 2) is nonplanar. $C_{5v}$ symmetry is more common, including the nonplanar conformation of corannulene **101,** which undergoes rapid dynamic movement [37], appropriately substituted derivatives of the elegant **102** synthetized by the Prinzbach group [38], and the average structure of the most symmetrical conformation of [5]calixarene **103** [39]. It is easy to construct a hypothetical molecule of $C_{6v}$ symmetry by appropriate substitution of molecules of $C_{6h}$ or $D_{6h}$ symmetry shown below. However, to our knowledge no

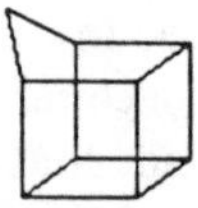

**92**

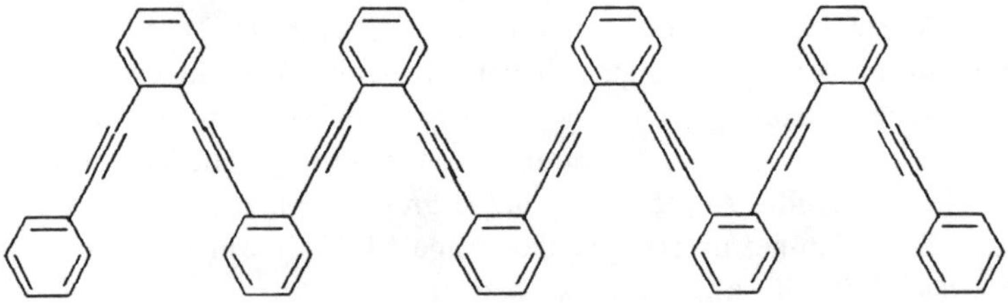

**93**

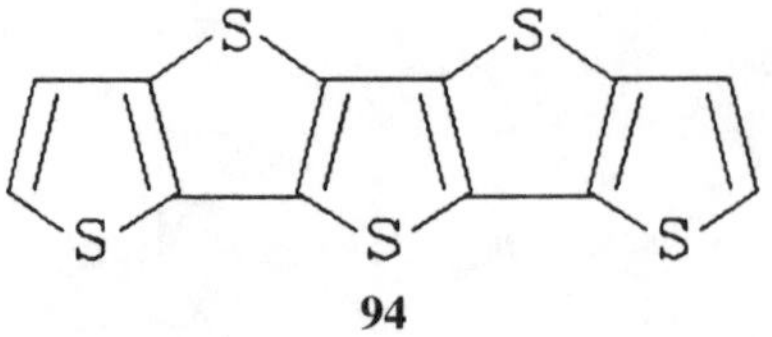

**94**

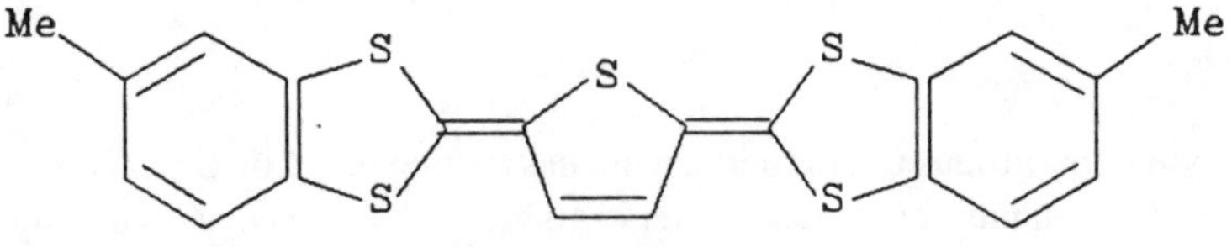

**95**

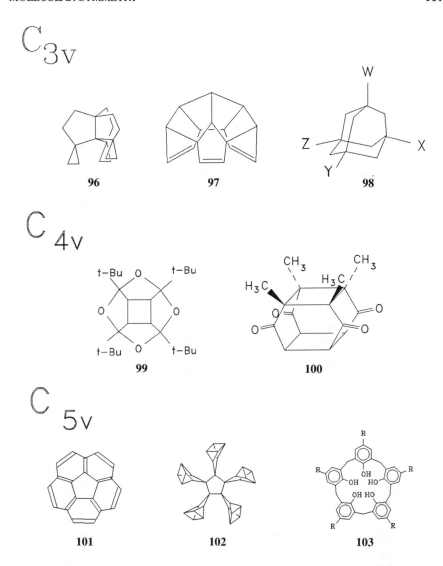

$C_{3v}$

**96**          **97**          **98**

$C_{4v}$

**99**          **100**

$C_{5v}$

**101**          **102**          **103**

such molecule is known. The same is true for higher symmetries with finite $n$ values. All linear molecules with no additional symmetry elements, such as HC≡N **43** ($n = 1$) (discussed in Chapter 2) [40], HC≡CCl, etc., belong to the $C_{\infty v}$ group.

Symmetry groups $C_{nh}$ have a plane $\sigma_h$ (h stands for horizontal) perpendicular to the symmetry axis $C_n$. Molecules belonging to this group have one proper axis $C_n$, one symmetry plane $\sigma_h$, and one improper axis $S_n$. $C_{1h}$ is, of course, identical with the $C_s$ group. Some examples of molecules with the $C_{nh}$ symmetry are given in the following. *Trans*-disubstituted ethylene and **104–106** serve as examples of the molecules belonging to the $C_{2h}$ symmetry group [41–43]. **107** is not rigid, and the tips of its rings change their orientations, but the depicted conformation belongs

# $C_{2h}$

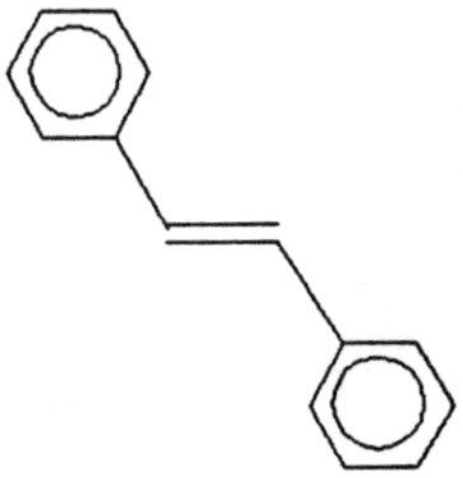

**104** 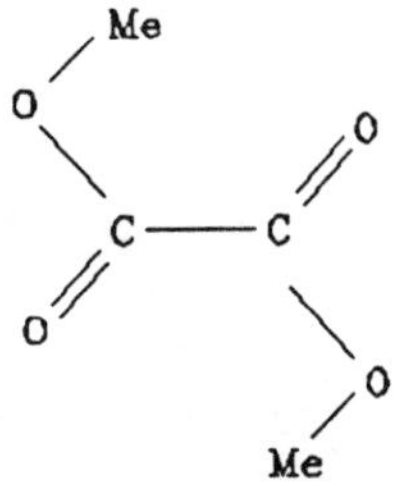                    **105** 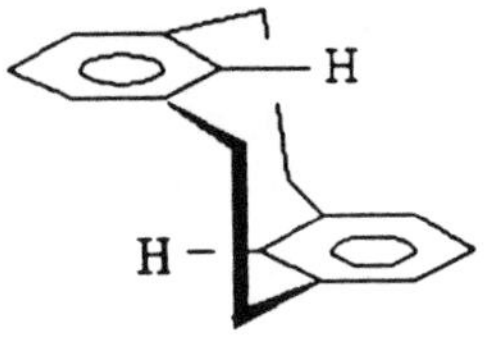                    **106**

# $C_{3h}$

**107**

to the $C_{3h}$ symmetry group [44]. If a central ring in tetrachloro-[4]-radialene **108** [45] were planar, then the molecule would have $C_{4h}$ symmetry. Analogous derivatives of [5]-radialene having $C_{5h}$ symmetry seem not to be known [46]. If the central ring in analogous [6]radialene were planar, then this molecule would belong to the $C_{6h}$ symmetry group. (See, however, the discussion of the structure of the hexasulfur analog $C_6S_6$ **49** of [6]radialene in chapter 2.) Heptacyclic hexalactame **109** [47] also belongs to this group. An interesting reaction leading to the latter molecule is presented in Section 10.5.

Symmetry groups $D_n$: If in addition to a $C_n$ axis a molecule has $n$ $C_2$ axes perpendicular to it, then the molecule belongs to the $D_n$ symmetry group. The group $D_2$ consists of three mutually perpendicular $C_2$ axes, while in the $D_3$ one there is a $C_3$ axis and three $C_{2n}$ axes perpendicular to the first one. Vespirenes **110** [48], betweenanene **111** [49], and twistane **112** [50], as well as the chiral isomer of the McCasland quaternary ammonium salt **75b** discussed earlier, have $D_2$ symmetry. As will be discussed in detail in the next chapter, the study of symmetry and chirality of all isomers of this salt enabled the formulation of the symmetry-based condition of chirality. This has ended a century of controversy, since for a long time the presence

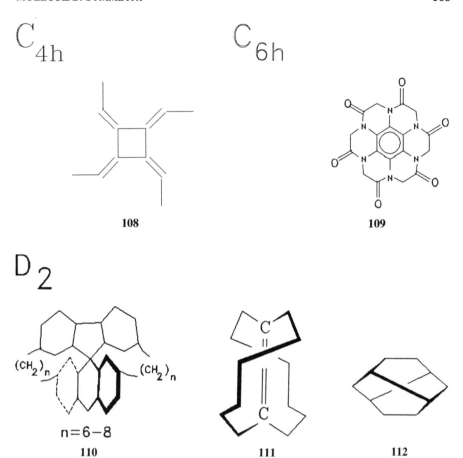

$C_{4h}$

$C_{6h}$

108

109

$D_2$

$(CH_2)_n$ $(CH_2)_n$

$n=6-8$

110

111

112

of an asymmetric atom in a molecule was thought to be a necessary and sufficient condition of molecular chirality. *trans,trans,trans*-perhydrotriphenylene **113** synthesized by Farina [51], chiral trishomocubane **114** [52], and a nonplanar conformation of **115** [53] have $D_3$ symmetry, while the average structure of **116** has a rare $D_4$ symmetry [54]. Molecules **113** and **114** have been resolved into enantiomers.

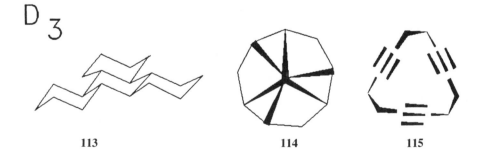

$D_3$

113

114

115

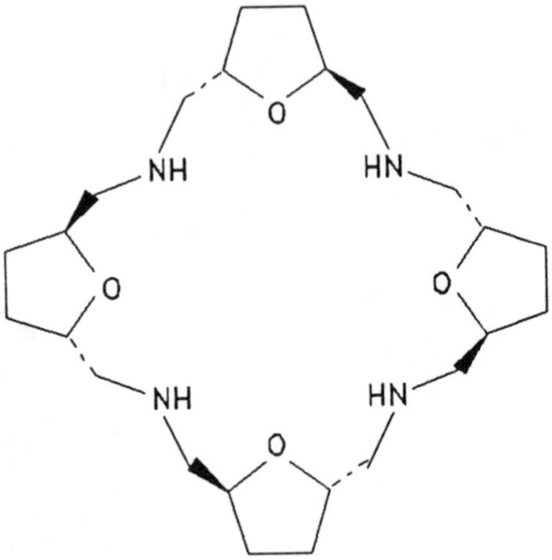

**116**

$\mathbf{D_{nd}}$ symmetry groups: If in addition to a $C_n$ axis and perpendicular to it $n$ $C_2$ axes a molecule has $n$ symmetry planes $\sigma$ intersecting at the $C_n$ axis, then this molecule belongs to the $\mathbf{D_{nd}}$ symmetry group. Such molecules also have improper symmetry axes $S_{2n}$ overlapping with the $C_n$ axis and, for odd values of $n$, a symmetry center $i$. Allene **117** [55] and spiropentane **118** [56] have $\mathbf{D_{2d}}$ symmetry, while cyclohexane **22** and its hexasubstituted derivatives **119,** as well as ypticene **120** [57] in a conformation with appropriately twisted [2.2.2]bicyclooctane units and bicubyl **121** [58], have $\mathbf{D_{3d}}$ symmetry.

$\mathbf{D_{2d}}$

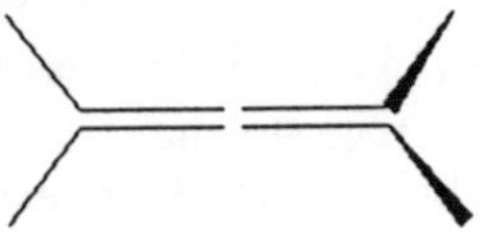

**117**

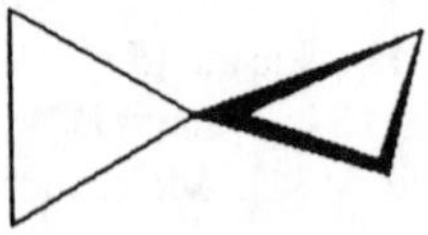

**118**

$\mathbf{D_{3d}}$

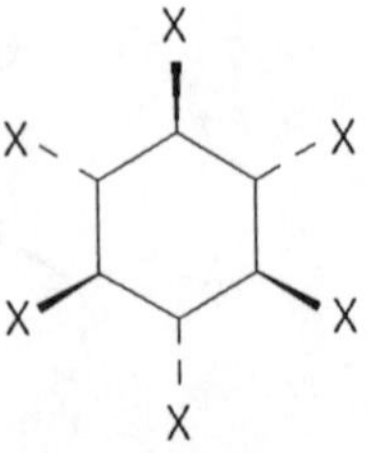

**119**

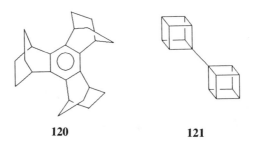

**120**        **121**

$\mathbf{D_{nh}}$ symmetry groups: If in a molecule in addition to a $C_n$ symmetry axis and $n$ symmetry planes $\sigma_v$ intersecting at the $C_n$ there is a symmetry plane $\sigma_h$ perpendicular to the axis, then this molecule belongs to a $\mathbf{D_{nh}}$ symmetry group. Some recently synthesized elegant molecules of these symmetries will be shown here. Pagodane **122** [59]; [3]rotane **71** [60], for which disagreement between the experimental results and molecular mechanics calculations forced a reinvestigation of its geometry, yielding a correction of the experimental data (discussed in Section 3.3); as well as benzene and the molecules **123, 124,** and **127,** [61, 62]; the average structures of **125** [63], **126** [64], and that of **128** [65] have $\mathbf{D_{nh}}$ symmetry. **129** [66] and **130** [67] would also have $\mathbf{D_{6h}}$ symmetry if their central rings were planar. The aromaticity of the latter molecule was discussed recently by Cioslowski et al. [68]. Linear mole-

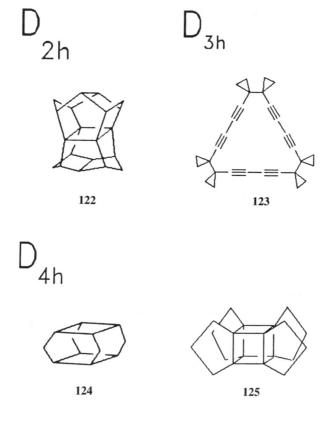

$D_{2h}$            $D_{3h}$

**122**            **123**

$D_{4h}$

**124**            **125**

# D$_{5h}$

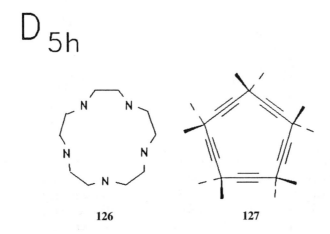

126                                        127

# D$_{6h}$

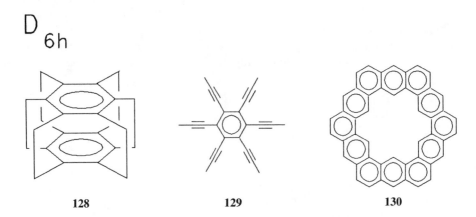

128                      129                      130

cules with a symmetry plane perpendicular to the molecular axis HC≡CH, $CO_2$, and cyanopolyynes NC—$(C≡C)_n$—CN ($n$ = 3–7) [69] have $D_{∞h}$ symmetry.

**T, T$_d$, O, O$_h$,** and **I$_h$** symmetry groups: In a molecule there can be more than one symmetry axis of order higher than two. For instance, four $C_3$ axes and three perpendicular $C_2$ axes together with the identity element $E$ form the group **T**. According to the calculations carried out by Iroff and Mislow [70], the hypothetical tetra-t-butylmethane $C(C(CH_3)_3)_4$ **131** has twisted t-butyl groups. In such a conformation it belongs to the group **T**. Substituting the tertiary hydrogen atoms $X, Y, Z, W$ of adamantane **98** by a group of $C_3$ symmetry should also yield a molecule of **T** symmetry while analogous substitution involving trishomocubyl moiety produced a chiral molecule of **D$_3$** symmetry [71]. The depicted conformation of the appealing cage compound **132** [72] also has this symmetry.

In the **T$_d$** symmetry group, molecules such as methane (see, however, Chapter 2 on MW spectra and the dipole moment of methane in the ground electronic and

T

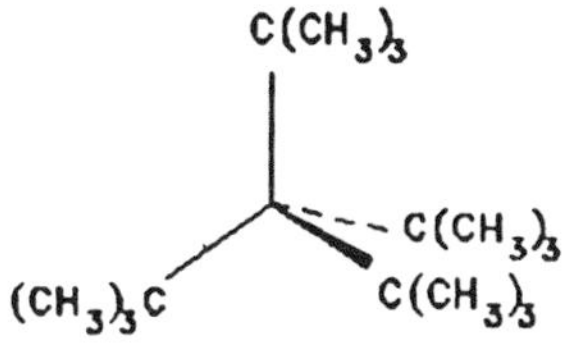

131

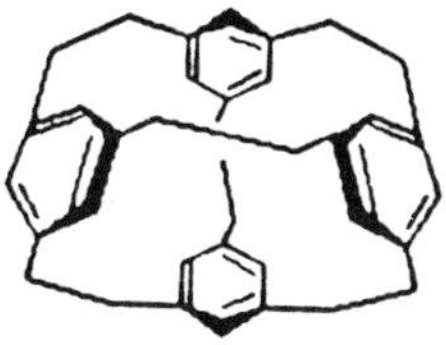

132

$T_d$

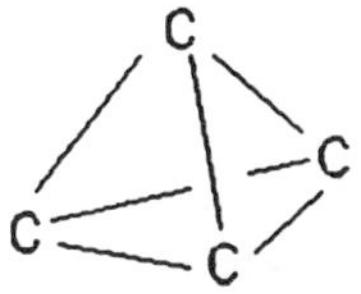

133

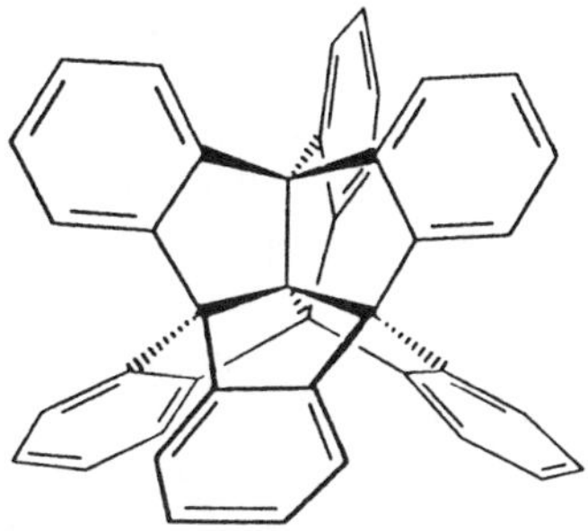

134

vibrational states, which display very small distortions from this high-symmetry structure), adamantane **98** with four identical substituents ($X = Y = Z = W$), as well as hypothetical tetrahedrane **133** [73] (discussed in Section 7.2.4.1), and elegant centrohexaindane **134** synthesized by Kuck [74] have, in addition to the symmetry elements of the **T** group six symmetry planes $\sigma$ and three $S_4$ axes overlapping with $C_2$ axes.

If a molecule has three mutually perpendicular $C_4$ axes and four $C_3$ axes, which have the same orientations with respect to one another as the $C_2$ and $C_3$ axes of the point group **T**, it belongs to the octahedral point group **O**. If in addition to the elements of symmetry of the octahedral group there is a center of symmetry $i$ in the molecule under study, then the molecule belongs to the cubic symmetry group **$O_h$**. This molecule is, of course, cubane **7** [75]. Inorganic borane **135** and sulfur hexafluoride **136** also belong to this symmetry group. The regular pentagon dodecahedron **8** [76] and icosahedron **11** discussed in Chapter 8 belong to the point group **$I_h$**. In his comprehensive overview of molecular symmetry groups, Herzberg, later awarded with a Nobel prize, stated in 1945 [77] that "It is not likely that molecules of such a symmetry will ever be found." He had to change this opinion in less than 40 years, providing another example of the validity of the proverb "You should never say never."

$O_h$

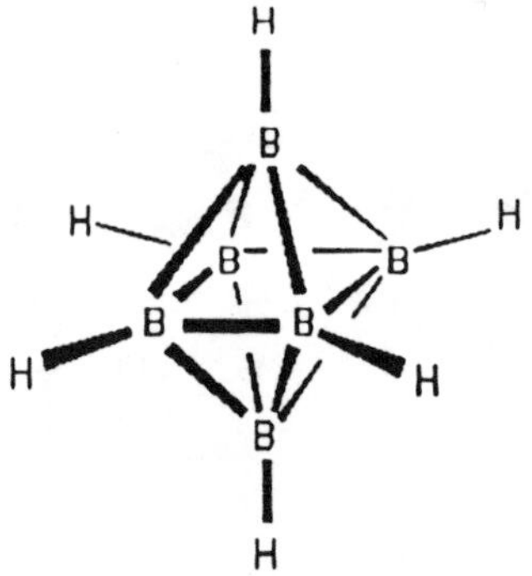

135 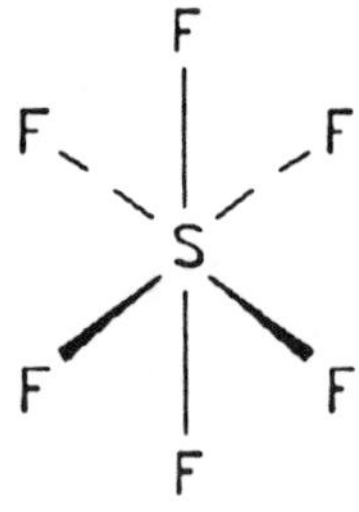                                              136

It should also be stressed that molecular formulae do not always show molecular symmetry correctly. Radialenes are probably the best examples of such a misleading presentation. For instance, due to steric crowding, the beautiful **137** [78] has no fivefold axis since the central ring is nonplanar.

Molecular point symmetry groups with their most important symmetry elements are summarized in Table 4.1.

It would probably be instructive for a reader less familiar with molecular symmetry to build models and follow the changes of methane and ethylene symmetry upon subsequent substitutions as presented in Figs. 4.3 and 4.4, respectively.

As discussed before, the analysis of symmetry properties of molecules that exist as conformer mixtures depends on the physicochemical method applied for the observation. If the rate of interconversion between the conformers is faster than the *time constant* of the method, only the averaged structure is observed. For instance, it is well known that cyclohexane at room temperature is a mixture of two rapidly interconverting chair forms. The process is called *ring inversion*. When analyzing

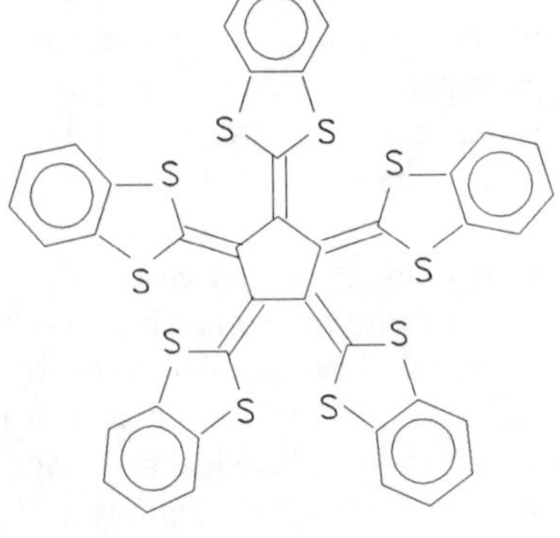

137

**Table 4.1** Point Groups and Their Main Symmetry Elements

| | |
|---|---|
| $\mathbf{C_1}$ | No symmetry |
| $\mathbf{C_n}$ | $n$-fold symmetry axis $C_n$ |
| $\mathbf{C_i = S_2}$ | Inversion center $i$ |
| $\mathbf{C_S = C_{1v} = C_{1h}}$ | Symmetry plane $\sigma$ |
| $\mathbf{S_n}$ | Improper $n$-fold symmetry axis; for even $n$ $S_n \equiv C_n + i$ |
| $\mathbf{C_{nv}}$ | One $C_n$ axis and $n$ planes $\sigma_v$; their intersection coincides with the axis |
| $\mathbf{C_{\infty v}}$ | One $C_\infty$ axis and an infinite number of $\sigma_v$ planes |
| $\mathbf{C_{nh}}$ | One $C_n$ axis and the plane $\sigma_h$ perpendicular to it |
| $\mathbf{D_n}$ | One $C_n$ axis and $n$ $C_2$ axes perpendicular to it |
| $\mathbf{D_{nd}}$ | One $C_n$ axis and $n$ $C_2$ axes perpendicular to it plus $n$ planes $\sigma_v$ and an $S_{2n}$ axis |
| $\mathbf{D_{nh}}$ | One $C_n$ axis and $n$ $C_2$ axes perpendicular to it plus $n$ planes $\sigma_v$ and the $\sigma_h$ plane; for even $n$ also an $S_n$ axis |
| $\mathbf{D_{\infty h}}$ | One $C_\infty$ axis and perpendicular to it a $\sigma_h$ plane |
| $\mathbf{T}$ | Three mutually perpendicular $C_2$ axes; four $C_3$ axes |
| $\mathbf{T_d}$ | Three mutually perpendicular $C_2$ axes, four $C_3$ axes, six $\sigma$ planes, three $S_4$ axes coinciding with $C_2$ axes |
| $\mathbf{O}$ | Three $C_4$, four $C_3$, and six $C_2$ axes |
| $\mathbf{O_h}$ | Three $C_4$, four $C_3$, and six $C_2$ axes, $i$, six $S_4$ improper axes coinciding with $C_2$, nine planes $\sigma$, four $S_6$ coinciding with $C_3$ |
| $\mathbf{I}$ | Six $C_5$, ten $C_3$, and 15 $C_2$ axes |
| $\mathbf{I_h}$ | Six $C_5$, ten $C_3$, and 15 $C_2$ axes, and $i$ |

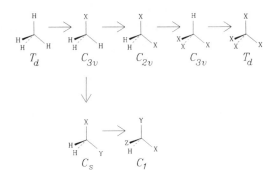

**Figure 4.3**  Changes in methane symmetry upon substitution.

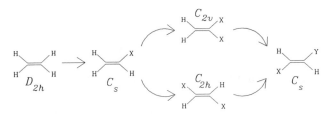

**Figure 4.4**  Changes in ethylene symmetry upon substitution.

experimental results for the molecule, one has to establish whether an experimental technique "sees" a chair structure of $D_{3d}$ symmetry characterized by two distinct hydrogen atom positions or the molecule behaves as if it had an averaged structure with $D_{6h}$ symmetry. Similarly, the averaged structure of *cis*-decalin **25** has $C_{2v}$ symmetry, while the symmetry is lowered to $C_2$ for its enantiomers with frozen inversion. The temperature-dependent $^{13}C$ NMR spectra of *cis*-decalin discussed in detail in Section 2.3.7.3 illustrate this point. Their study [79] allows one to determine not only the energy difference between two conformers (equal to zero as is in the case of cyclohexane conformers), but also the energy barrier between the conformers.

## References

1. K. Mislow and P. Bickart, *Isr. J. Chem.* **15** (1976/77) 1.

2. F. A. Cotton, *Chemical Applications of Group Theory*, Wiley, New York, 1990; I. Hargittai and M. Hargittai, *Symmetry through the Eyes of a Chemist*, VCH, Weinheim, 1986; E. P. Wigner, *Group Theory and Its Applications to the Quantum Mechanics of Atomic Spectra*, Academic Press, New York, 1959.

3. G. Herzberg, *Infrared and Raman spectra of Polyatomic Molecules*, Van Nostrand, New York, 1946, p. 1.

4. K. C. Ingham and S. J. Strickler, *J. Chem. Phys.* **53** (1970) 4313.

5a. Schönflies' notation is mostly used by chemists and spectroscopists.

5b. The Hermann–Maugin or International system (see J. D. Dunitz, *X-Ray Analysis and the Structure of Organic Molecules*, Cornell University Press, Ithaca, 1979, p. 76) is used by crystallographers.

6a. R. D. Johnson, G. Meier, and D. S. Bethune, *J. Am. Chem. Soc.* **112** (1990) 8983.

6b. C. S. Yannoni, P. P. Bernier, D. S. Bethune, G. Meijer, and J. R. Salem, *J. Am. Chem. Soc.* **113** (1991) 3190.

7. W. Huber and K. Möllen, *Acc. Chem. Res.* **19** (1986) 300.

8. Z. Yoshida and S. Miki, unpublished results cited in H. Dodziuk, D. Sybilska, S. Miki, Z. Yoshida, J. Sitkowski, M. Asztemborska, A. Bielejewska, J. Kowalczyk, K. Duszczyk, and L. Stefaniak, *Tetrahedron*, **50** (1994) 3619.

9. G. E. McCasland and S. Proskow, *J. Am. Chem. Soc.* **77** (1955) 4688; **78** (1956) 5646.

10. J.-M. Flaud, C. Camy-Peyret, J. W. C. Johns, and B. Carli, *J. Chem. Phys.* **91** (1989) 1504.

11. R. H. Martin, *Angew. Chem., Int. Ed. Engl.* **13** (1974) 649.

12. A. C. Cope, C. R. Ganellin, H. W. Johnson, T. V. van Auken, and H. J. S. Winkler, *J. Am. Chem. Soc.* **85** (1963) 3276.

13. K. Mislow, *Acc. Chem. Res.* **9** (1976) 26.

14. W. Spielman and A. de Meijere, *Angew. Chem.* **15** (1976) 429.

15. R. C. Fort, Jr., and W. L. Semon, *J. Org. Chem.* **32** (1967) 3685.

16. J. Furukawa, T. Kakuzen, H. Morikawa, R. Yamamoto, and O. Okuno, *Bull. Chem. Soc. Jpn.* **41** (1968) 155.

17. R. Gerdil, *Top. Curr. Chem.* **140** (1987) 71; G. A. Facey and J. A. Ripmeester, *J. Chem. Soc. Chem. Commun.* (1990) 1585; J. A. Ripmeester and N. W. Burlinson, *J. Am. Chem. Soc.* **107** (1985) 3713; R. Gerdil and G. Barchietto, *Tetrahedr. Lett.* **28** (1987) 4685; *ibid.* **30** (1989) 6677.

18. A. C. D. Newman and H. M. Powell, *J. Chem. Soc.* (1952) 3747.

19. Three atoms of NOCl determine a molecular plane, and there are no other additional symmetry elements. The consequences of its planarity on IR and Raman spectra are discussed in Ref. 3, p. 105.

20. D. Christen, H.-G. Mack, C. J. Marsden, H. Oberhammer, G. Schatte, K. Sppelt, and H. Wilner, *J. Am. Chem. Soc.* **109** (1987) 4009.

21. M. Nogradi, *Stereochemistry. Basic Concepts and Applications,* Akademiai Kiado, Budapest, 1981, p. 21.

22. F. A. L. Anet, *J. Am. Chem. Soc.* **84** (1962) 671.

23. L. Fitjer and U. Quabeck, *Angew. Chem., Int. Ed. Engl.* **26** (1987) 1023.

24. D. Wehle, N. Schormann, and L. Fitjer, *Chem. Ber.* **121** (1988) 2171.

25. D. A. Dixon, J. C. Calabrese, and J. S. Miller, *Angew. Chem., Int. Ed. Engl.* **28** (1989) 90.

26. Z. Goren and S. E. Biali, *J. Am. Chem. Soc.* **112** (1990) 893.

27. L. A. Paquette, J. S. Ward, R. A. Boggs, and W. B. Farnham, *J. Am. Chem. Soc.* **97** (1975) 1101.

28. F. Wudl and S. P. Bitler, *J. Am. Chem. Soc.* **108** (1986) 4685.

29. Y. Mazaki and K. Kobayashi, *Tetrahedr. Lett.* **30** (1989) 3315.

30. C. Rovira, N. Santalo, and J. Veciana, *Tetrahedr. Lett.* **30** (1989) 7249; K. Takahashi, T. Nihira, K. Takase, and K. Shibata, *Tetrahedr. Lett.* **30** (1989) 2091.

31. J. E. Maggio, H. E. Simmons III, and J. K. Kouba, *J. Am. Chem. Soc.* **103** (1981) 1579.

32. L. Paquette, R. A. Snow, J. L. Muthard, and T. Cynkowski, *J. Am. Chem. Soc.* **101** (1979) 6991.

33. R. C. Bingham and P. v. T. R. Schleyer, *Top. Curr. Chem.* **18** (1971) 1.

34. A. Frondo and G. Maas, *Angew. Chem. Int. Ed. Engl.* **28** (1989) 1663.

35. R. C. Cookson, D. A. Cox, and J. Hudec, *J. Chem. Soc. (London)* (1961) 4499.

36. H. Kämmerer, G. Happel, and F. Caesar, *Macromol. Chem.* **162** (1972) 179; G. Happel, B. Mathiasch, and H. Kaemmerer, *ibid.* **176** (975) 3317; J. H. Munch, *ibid.* **178** (1977) 69.

37. W. E. Barth and R. G. Lawton, *J. Am. Chem. Soc.* **93** (1971) 1730; A. H. Abdourazak, A. Sygula, and P. W. Rabideau, *J. Am. Chem. Soc.* **115** (1993) 3010.

38. S. Trah, K. Weidmann, H. Fritz, and H. Prinzbach, *Tetrahedron Lett.* **28** (1987) 4399.

39. Ref. 36.

40. Ref. 3, p. 279.

41. M. Traetteberg, E. B. Frantzen, F. C. Mijlhof, and A. Hoekstra, *J. Mol. Struct.* **26** (1975) 69.

42. B. P. van Eijck and F. B. Duijneveldt, *J. Mol. Struct.* **39** (1977) 2, 157.

43. E. Müller and G. Röscheisen, *Chem. Ber.* **90** (1957) 543.

44. M. Doyle, W. Parker, and P. A. Gunn, *Tetrahedron Lett.* (1970) 3619; J. C. Coll, D. R. Crist, M. C. G. Barrio, and N. J. Leonard, *J. Am. Chem. Soc.* **94** (1972) 7092.

45. H. Hoppf and A. K. Wick, *Helv. Chim. Acta* **44** (1961) 19; *ibid.* **46** (1961) 380.

46. H. Hopf and G. Maas, *Angew. Chem., Int. Ed. Engl.* **31** (1992) 931.

47. J. Thomaides, P. Maslak, and R. Breslow, *J. Am. Chem. Soc.* **109** (1986) 3970.

48. G. Haas and V. Prelog, *Helv. Chim. Acta* **52** (1969) 1202.

49. M. Nakazaki, K. Yamamoto, and J. Inagi, *J. Chem. Soc. Chem. Commun.* (1977) 346; M. Nakazaki, K. Yamamoto, and J. Inagi, *J. Am. Chem. Soc.* **101** (1979) 147.

50. H. W. Whitlock, Jr., *J. Am. Chem. Soc.* **90** (1968) 4929.

51. M. Farina and G. Audisio, *Tetrahedron Lett.* (1967) 1285; *Tetrahedron* **26** (1970) 1839.

52. M. Nakazaki, K. Naemura, N. Arashiba, and M. Iwasaki, *J. Org. Chem.* **44** (1979) 2433.

53. G. Wilke and M. Kröner, *Angew. Chem.* **71** (1959) 574; G. Allegra and I. W. Bassi, *Atti Accad. Naz. Linzei* **38** (1962) 72.

54. K. Naemura, Y. Hokura, Y. Kanda, and M. Nakazaki, *Chem. Lett.* (1985) 615.

55. F. Takabayashi, M. Sugie, and K. Kuchitsu, Seventh Austin Symposium on Gas Phase Molecular Structure, Austin, Texas, 1978, p. 103, cited in L. V. Vilkov, V. S. Mastriukov, and N. I. Sadova, *Determination of the Geometrical Structure of Free Molecules,* p. 84.

56. R. Boese, D. Blaeser, K. Gomann, and V. H. Brinker, *J. Am. Chem. Soc.* **111** (1989) 1501.

57. H. Hart, A. Bashir-Hashemi, J. Luo, and M. A. Meador, *Tetrahedron* **42** (1986) 1641.

58. P. E. Eaton and M. Maggini, *J. Am. Chem. Soc.* **110** (1988) 7230.

59. W. D. Fessner, G. Sedelmeier, P. R. Spurr, G. Rihs, and H. Prinzbach, *J. Am. Chem. Soc.* **109** (1987) 4626.

60. R. Boese, T. Miebach, and A. de Meijere, *J. Am. Chem. Soc.* **113** (1991) 1743.

61. A. de Meijere, F. Jaekel, A. Simon, H. Borrmann, J. Köhler, D. Johnels, and L. T. Scott, *J. Am. Chem. Soc.* **113** (1991) 3935.

62. R. Gleiter and M. Karcher, *Angew. Chem., Int. Ed. Engl.* **27** (1988) 840.

63. R. J. Gelb, J. S. Alper, and M. H. Schwartz, *J. Phys. Org. Chem.* **5** (1992) 443.

64. Ref. 61.

65. R. Diercks, J. C. Armstrong, R. Boese, and K. P. C. Vollhardt, *Angew. Chem., Int. Ed. Engl.* **25** (1986) 268; R. Diercks and K. P. C. Vollhardt, *ibid.* **25** (1986) 266.

66. F. Gerson, *Angew. Chem.* **95** (1983) 571.

67. H. A. Staab and F. Diederich, *Chem. Ber.* **116** (1983) 3487.

68. Cioslowski, P. B. O'Connor, and E. D. Fleischmann, *J. Am. Chem. Soc.* **113** (1991) 1086.

69. T. Grösser and A. Hirsch, *Angew. Chem., Int. Ed. Engl.* **32** (1993) 1340.

70. L. D. Iroff and K. Mislow, *J. Am. Chem. Soc.* **100** (1978) 2121.

71. K. Naemura, Y. Hokura, and M. Nakazaki, *Tetrahedron* **42** (1986) 1763.

72. F. Vögtle, J. Gross, C. Seel, and M. Nieger, *Angew. Chem., Int. Ed. Engl.* **31** (1992) 1069.

73. G. Maier, *Angew. Chem.* **100** (1988) 317.

74. D. Kuck, *Angew. Chem., Int. Ed. Engl.* **27** (1988) 1192.

75. P. E. Eaton and T. W. Cole, Jr., *J. Am. Chem. Soc.* **86** (1964) 962, 3157.

76. R. J. Ternansky, D. W. Balogh, and L. Paquette, *J. Am. Chem. Soc.* **104** (1982) 4503.

77. Ref. 3a, p. 12.

78. Z. Yoshida and T. Sugimoto, *Angew. Chem. Adv. Mater.* **27** (1988) 1573.

79. D. K. Dalling, D. M. Grant, L. F. Johnson, *J. Am. Chem. Soc.* **93** (1971) 3678; B. E. Mann, *J. Magn. Reson,* **21** (1976) 17.

# 5

# Chirality

## 5.1 The Importance of Chirality in Nature

The word *chirality* is derived from the Greek χειροσ, meaning hand. It describes the properties of an object that cannot be superimposed on its mirror image (see, for instance color chart 3, presenting chloro-, bromo-, and iodomethane in both arrangements). The importance of chirality cannot be overestimated, since the prevailing majority of organic substances from which all living creatures are built is chiral. Sugars, proteins, and nucleic acids, as well as the majority of simpler organic molecules, are chiral, and they are produced in nature with unrivaled *stereoselectivity*. To name only a few examples of the stereo- and *enantioselectivity*:

1. Until recently, only L-amino acids, like L-serine **138a,** were thought to be produced by living creatures [1]. Today there is growing evidence of the presence of D-amino acids (like **138b**) not only in simple microorganisms, but even in mammals [2]. However, natural L-amino acids certainly constitute the prevailing majority. The questions "Why is life based on L-amino acids and D-sugars? . . ." [3] and "Why is nature so enantiospecific?" cannot be answered with certainty. Some calculations [3] indicate that it can be the result of weak nuclear forces, making one enantiomer very slightly more stable than the other.
2. Cholesterol **139** (the stars in the formula denote asymmetric carbon atoms to be discussed) clearly illustrates the striking stereospecificity of chemical reactions in nature. The molecule has 256 different chiral stereoisomers (see the discussion to follow). Its isomer cholest-5-en-3β-ol **139** is present in human organisms, while others are sometimes found in plants and marine organisms [4].

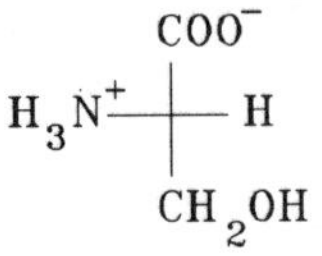

138a 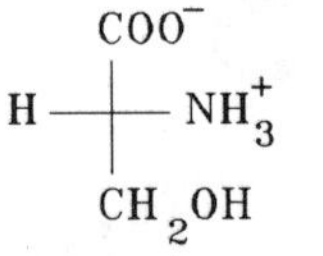          138b 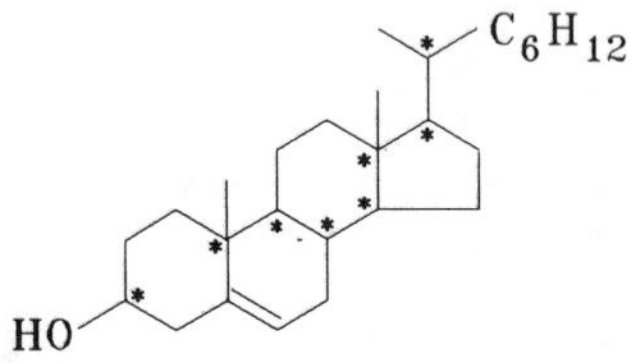          139

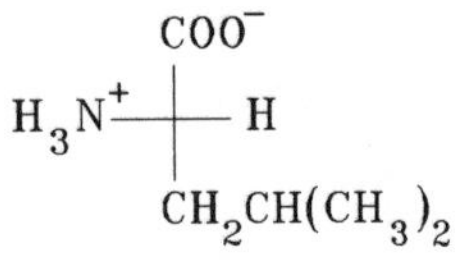

140a 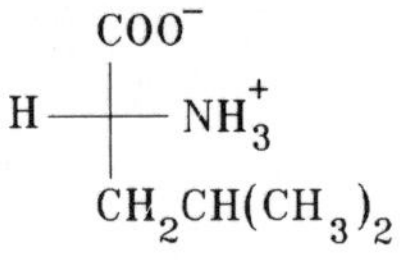          140b 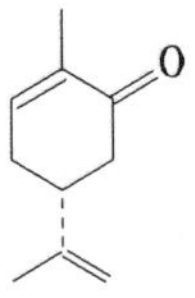          141a 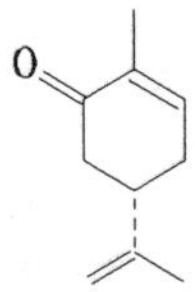          141b

3. Similarly, only L-amino acids such as L-leucine **140a** are produced by certain shells. When they die, their chiral constituent molecules very slowly racemize, and the amount of the second enantiomer in fossils depends on their age. This was established, for example, by an investigation of the content of D-amino acids in inner layers of Mercenaria shells of various age [5]. For instance, no D-leucine **140b** was found in the contemporary shells, while the amount in late Pleistocene shells (1 million to 10 thousand years old) was 26%. It increased to 48% in the much older Miocene shells (28 to 12 million years old). The speed of the racemization process depends on several factors, such as the acidity of the environment, temperature, etc. Therefore, it is not possible to base a determination of the age of organic fossils solely on the basis of the D-amino acid content.

4. A highly efficient chiral recognition is illustrated by the different odors we experience inhaling (+)- or (−)-carvone **141a** and **141b,** respectively. Both compounds produce an herbaceous odor, but the former is reminiscent of caraway and dill seeds, while the latter is suggestive of spearmint [6].

5. The phenomenon of chirality is of fundamental significance. However, it also has many practical consequences. Due to the high specificity of organic reactions in nature, the pharmacological activity of a drug depends on the sense of its chirality. For instance, the reaction of (R)- and (S)-nicotine **142** is highly enantiospecific [7]. As will be discussed later in this chapter, the consequences of the use of an improper enantiomer of a drug, even as an admixture, can be serious.

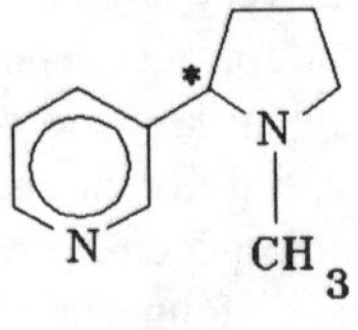

142

There is no uniformity in notations of molecular configuration. Three systems are used to denote the sense of molecular chirality (which reflects the immaturity of this domain). The first, D-,L-, dating back to Emil Fischer [8], denotes the configuration on the C2 carbon atom in amino acids (and that on the asymmetric carbon atom bearing the highest number in sugars). We believe that, for the sake of consistency, this system should be abandoned in favor of the Cahn, Ingold, and Prelog R,S notation [9] discussed later in this chapter. Here the consistency means that (1) the chirality of amino acids should be denoted in the same way as that of other organic compounds and (2) the same convention should be used to denote chirality of both asymmetric carbon atoms in L-*iso*-leucine **143** and L-threonine **144**. The second notation, (+) or (−), describes the sign of the rotation of the plane of polarization for the sodium *D* line in CD as discussed in Section 2.3.6. We believe that this system of notation should also be abandoned in favor of the R/S and P/M system in all cases when the absolute configuration is known. Otherwise, it should be preserved until the absolute configuration is determined. The third notation, making use of the R/S or P/M symbols, will be discussed in detail.

$$
\begin{array}{cc}
\text{COO}^- & \text{COO}^- \\
\text{H}_3\text{N}^+ \!\!-\!\!\!\!\mid\!\!-\!\! \text{H} & \text{H}_3\text{N}^+ \!\!-\!\!\!\!\mid\!\!-\!\! \text{H} \\
\text{H}_3\text{C} \!\!-\!\!\!\!\mid\!\!-\!\! \text{H} & \text{H} \!\!-\!\!\!\!\mid\!\!-\!\! \text{OH} \\
\text{C}_2\text{H}_5 & \text{CH}_3 \\
\mathbf{143} & \mathbf{144}
\end{array}
$$

In the following discussion the difference between a real chiral molecule and its model representation discussed in detail by Mislow and Bickart [10] has to be kept in mind. In particular, as previously, a molecule is considered to be a rigid set of point masses connected by bonds. Thus we shall analyze either a molecular structure in a frozen conformation or the mean averaged over intramolecular motion.

The simplest chiral molecule consists of an *asymmetric carbon atom,* that is, the carbon atom with its four different substituents. If a molecule possesses two or more asymmetric atoms, then a more complicated case of *stereoisomerism* arises. For instance, there is a possibility of at most four stereoisomers (i.e., two diastereomeric pairs of enantiomers) for two asymmetric atoms in a molecule. The classification, notation, and enumeration of stereoisomers involving chirality is the purpose of the classification of chirality [11]. Let us recall that two molecules that are mirror images of each other are called *enantiomers,* while stereoisomers that are not enantiomers are called *diastereoisomers.* As a result of nonstereospecific reactions a fifty–fifty mixture of two enantiomers, called a *racemate,* is obtained. Without a differentiating chiral factor, such as polarized light or another interacting chiral molecule, the enantiomers are practically identical [3]; that is, they have the same energy, bond lengths and angles, etc. Remembering that a hand is the most demonstrative model of a chiral object, one can easily see that two enantiomeric molecules visualized by the right and left hands interact in a different way than two molecules of the same chirality, such as two right hands. Such differences are of utmost significance in many domains, in particular in pharmacology, since in the case of a chiral drug a desired biological activity has, as a rule, only one enantiomer.

If, as is often the case, the drug is administered as the racemate, the second enantiomer constitutes a 50% impurity, which, at best, is a ballast. In the worst case it can cause serious side effects. The thalidomide (contergane) **145** tragedy [12a] is probably the best-known case of this kind [12b]. The deformities of the babies born after their mothers had taken the drug was believed to result from the teratogenic activity of the second enantiomer **145b** present in the racemic drug. The thalidomide drama happened about 30 years ago, but at present most drugs are still marketed as racemates. Thus the availability of pure enantiomers is of great practical importance. The enantiomers or mixtures with an excess of one enantiomer can be obtained either by asymmetric synthesis or by total or partial resolution of racemic mixtures. Today, asymmetric synthesis is a rapidly developing field. However, we are far from matching the amazing stereospecificity of organic reactions carried out in living organisms. Therefore, the separation of enantiomers and diastereoisomers (and more generally chirality) is a hot topic.

**145a**                    **145b**

Teratogenic S–thalidomide

Chirality studies were made possible by Arago's works on the optical activity of crystals [13] and Biot's observation that certain organic molecules in solution or in the gas phase rotate the plane of polarized light [14]. Then Pasteur's pioneering works on the isomers of tartaric acid **1–3** [15] led to the foundation of this field. All modern methods of resolution date back to his work. They consist of (1) mechanical separation of spontaneously crystallizing enantiomorphic crystals, (2) resolution through the formation of diastereoisomers, either complexed or covalently bound, and (3) resolution by chromatography. One of the best separation methods is chromatography using cyclodextrins (discussed in some detail in Chapter 10) as stationary or mobile chiral phases. It may be recalled that, in addition to his talent and diligence, Pasteur also had a bit of luck in these studies. He noticed that different crystals are formed by the precipitation of sodium ammonium tartrate and separated the enantiomers **1** and **2** manually. Happily, he carried out his experiments below 300 K, since at higher temperatures the compound crystallizes in the racemic form containing an equal number of (+) and (−) isomers.

As discussed in Section 2.3.7, ORD and CD methods are commonly used to detect optical activity, which manifests the chirality of the molecule under study. However, the only method yielding direct information on absolute configuration of the molecule under consideration is the study of anomalous X-ray scattering discussed in Section 2.2.1.2.

In spite of a great amount of experimental data, the field of chirality is still developing. After the formulation of van't Hoff [16] and LeBel's [17] hypothesis on the bond directionality of a tetravalent carbon atom, the presence of an asymmetric atom in a molecule was thought to be the condition of its chirality. Then new chiral molecules without such an atom were obtained. As mentioned in the preceding chapter, the synthesis of all four isomers of spiro ammonium salts **75a–d** played a decisive role in the formulation of a general symmetry-based condition of chirality [18] since it turned out that the isomer **75a** possessing an improper $S_4$ axis was not chiral. (Remember that an improper symmetry axis $S_n$ corresponds to a superposition of a rotation by an angle of $360°/n$ around the axis with the reflection in the plane perpendicular to this axis.) Thus the *chirality condition* states that a molecule is chiral if it does not have improper symmetry axes $S_n$. As discussed in Chapter 4, the $S_1$ axis is equivalent to a symmetry plane. Similarly, $S_2$ is equivalent to an inversion center. Therefore, molecules possessing an inversion center or symmetry planes (belonging to $\mathbf{C_{nv}}$, $\mathbf{C_{nh}}$, $\mathbf{D_{nd}}$, $\mathbf{D_{nh}}$, $\mathbf{T_d}$, $\mathbf{O_h}$, and $\mathbf{I_h}$) are achiral. On the other hand, the presence of $C_n$ axes does not preclude chirality, and the molecules belonging to $\mathbf{C_n}$, $\mathbf{D_n}$, $\mathbf{T}$, $\mathbf{O}$, and $\mathbf{I}$ symmetry groups are chiral.

## 5.2 The Cahn, Ingold, and Prelog Classification of Chirality, Its Merits, and Limitations

More than 30 years ago Cahn, Ingold, and Prelog published their famous classification of chirality [9], in which they named three elements of chirality: a center, an axis, and a plane of chirality, and further gave helicity, that is, the chirality of helicenes (such as **76**) and DNA spirals, and the conformational chirality of the *gauche* butane **28b** and **c** type as additional kinds of chirality. It should be stressed that these elements have been exemplified rather than precisely defined. Thus an asymmetric carbon atom **146** served as an example of the center of chirality, and appropriately substituted allenes **147**, alkylidenecyclohexanes **148**, spiranes **149**, adamantanes **150**, and biphenyls **29** were named as possessing axes of chirality and substituted cyclophanes **151** as molecules exhibiting a plane of chirality. Although a center of chirality was analyzed almost exclusively in terms of an asymmetric carbon atom, the possibility of the center not coinciding with any atom was mentioned. Two systems of chirality descriptors, R/S and P/M, have been proposed, which should be equivalent and interchangeable in certain cases for which both could be applied.

To assign the R or S symbol to an asymmetric atom, one has to determine priority of its substituents, making use of the ordering given in Table 5.1. To carry out the assignment, the model of the molecule is oriented in such a way (Fig. 5.1) that the

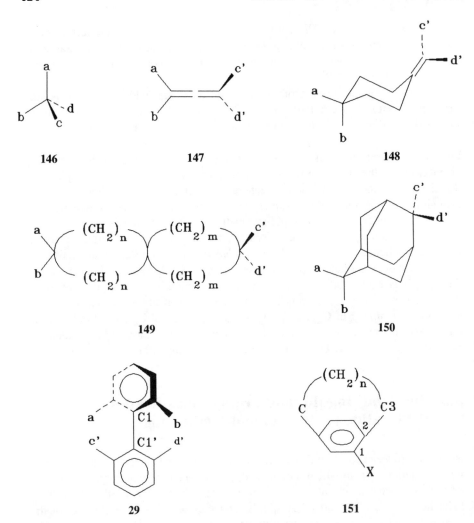

146                         147                              148

149                                              150

29                                              151

**Table 5.1**  The Increasing Order of Precedence of Atoms and
              Groups in the CIP Classification

| 1. H | 13. $C_6H_5$ | 25. OH |
|------|--------------|--------|
| 2. D | 14. $CH_2OH$ | 26. $OCH_3$ |
| 3. $CH_3$ | 15. $CH{=}O$ | 27. $OC_6H_5$ |
| 4. $CH_2CH_3$ | 16. $COR$ | 28. $OCOR$ |
| 5. $CH_2(CH_2)_nCH_3$ | 17. $CONH_2$ | 29. F |
| 6. $CH_2{-}CH{=}CH_2$ | 18. COOH | 30. SH |
| 7. $CH_2{-}C{\equiv}CH$ | 19. $COOR$ | 31. $SR$ |
| 8. $CH_2C_6H_5$ | 20. $NH_2$ | 32. $SOR$ |
| 9. $CH(CH_3)_2$ | 21. $NHCH_3$ | 33. $SO_2R$ |
| 10. $CH{=}CH_2$ | 22. $N(CH_3)_2$ | 34. Cl |
| 11. $C(CH_3)_3$ | 23. NO | 35. Br |
| 12. $C{\equiv}CH$ | 24. $NO_2$ | 36. I |

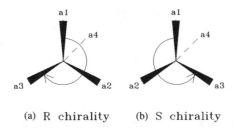

(a) R chirality    (b) S chirality

**Figure 5.1**   The way of determining R or S chirality of an asymmetric atom.

substituent with the lowest priority is located away from the observer. If the numbers a1, a2, a3 enumerating the priority of the substituents are arranged clockwise, then the atom has the R chirality. Otherwise, it is S.

The P/M designators show the sense of the screw formed by three consecutive vectors. As visualized in Fig. 5.2, the Newman projection (introduced in Chapter 1) of this molecular fragment has to be considered. Then the P (or M) designators represent the "+" (or "−") sign of the torsional angle formed by these vectors. It should be stressed that the P/M designators were applied in two different ways by Cahn, Ingold, and Prelog: They either described the mutual orientation of the substituents around a central bond of the *gauche* butane type, or they could refer to the overall chirality of secondary helical or propellerlike structures such as hexahelicene **76** or [4.4.4]propellane **152**. Although molecules consisting of atoms with coordination numbers less than or equal to four were the main target of the CIP classification, some rules concerning complexes with numbers equal to 5 and 6 have also been given.

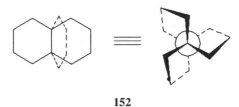

**152**

In 1982 Prelog and Helmchen [19] revised the classification. To define the elements of chirality they made use of the concept of a stereogenic unit. According to McCasland, an atom or a group of atoms represent a stereogenic unit if a permutation (that is, an interchange) of substituents produces a stereoisomer [20].

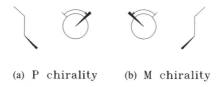

(a) P chirality    (b) M chirality

**Figure 5.2**   The way of determining P or M chirality of a helical axis.

In addition, Prelog and Helmchen summarized substituent priority rules, introducing minor modifications of the rules. They also formulated a recommendation prescribing the use of the R/S symbols exclusively for a description of the sense of chirality of centers of chirality and the P/M symbols solely for the description of the sense of axes and planes of chirality.

A well-known rule for diastereoisomer count states that for a molecule possessing $n$ asymmetric atoms the maximum number of stereoisomers due to the existence of these atoms is equal to $2^n$. (For this reason cholesterol **139**, mentioned before, having 8 asymmetric carbon atoms marked by stars, should have 256 stereoisomers associated with chirality.) This rule is of great practical importance since it states that there are at most two diastereomeric pairs of enantiomers for two asymmetric atoms in a molecule, etc. Eliel [22] suggested that this rule should be extended so that $n$ would include what was later called planes and axes of chirality. Some authors used the rule in this sense, but it seems that it has not been incorporated into textbooks, although it can be generalized.

A comparison of the Cahn, Ingold, and Prelog paper introducing the classification [9] with the Helmchen and Prelog article [19] illustrates the imperfection of the domain of chirality and a serious difficulty associated with it. Namely, not only do various authors have different ideas and use different concepts, but also works by the same authors are sometimes inconsistent. For instance, conformational chirality, introduced in Prelog's first paper, was not mentioned in the second one.

The CIP classification has been used in innumerable papers for isomer designation, but it was met with sharp criticism by several authors. The most serious critiques of the classification was formulated by Hirschmann and Hanson [21], Mislow and Siegel [11], and Dodziuk and Mirowicz [23]. The deficiencies of the CIP classification can be summarized in the following way:

1. The lack of definitions of the elements of chirality does not allow one to apply the classification to new topological molecules such as Walba's polyethers **30** [24] or knots **13** [25] discussed in Section 8.6.
2. Its application to some other systems is inconsistent. For instance, molecules **111** and **153** have the same $D_2$ symmetry, but according to the CIP classification, they have different elements of chirality. The former has a plane of chirality, while the latter has an axis of chirality.

$X = O, S, CO, CH_2$

**153**

3. To show that the elements of chirality cannot be defined in terms of a stereogenic unit, one should remember that stereogenicity and chirality are different properties that happen to overlap for the tetravalent carbon atom. The former property is assigned to atoms or groups of atoms for which a permutation of substituents produces a stereoisomer. The examples of stereogenic and achiral monosubstituted cyclohexane **154** and trimethoxymethane in the conformation **155** which is nonstereogenic and chiral, clearly show the difference between these properties. Therefore, the elements of chirality cannot be defined through stereogenic elements as demanded by Mislow and Siegel [11] and carried out by Prelog and Helmchen [19].

154                         155                         156

4. Conformational chirality has been introduced in an inconsistent way in the CIP classification. As discussed in Chapters 2 and 4, in this case one should also define and differentiate in each case whether a specific conformation or the averaged structure is considered. This question has not been precisely handled in the CIP classification. In its first version [9] for some molecules the chirality of specific conformations was considered. This is best represented by the cyclohexane **156** formula with P and M designators of the arrangement around each CC bond (the whole molecule is, of course, achiral). However, only the average structures of the parent unsubstituted biphenyl **35** (in which, as discussed in Section 2.2.1.4, in the gaseous state the aromatic rings are inclined at an angle of ca. 40°), as well as that of spiro[5.5]undecane **57a,b** [26], were taken into account in Refs. 9 and 19. Therefore, the chirality of **35** and **57a,b** has been overlooked in the classification.

   Let us analyze the first example of **35** and its derivatives **29** (Scheme 5.1) in some detail. Nonplanar biphenyl with frozen internal rotation has a $C_2$ symmetry axis; thus, it is chiral. If the rotation around the central $C_{sp^2}C_{sp^2}$ bond is fast, then only the average structure of **35** is detected by most experimental techniques, and the chirality of the molecule cannot be detected. The oversight of the conformational chirality of **35** and **57a,b** was rooted in the poor experimental feasibility of conformational isomers 30 years ago, but it resulted in an underestimation of conformational chirality. As shown in Section 2.3.8.3, one can monitor the manifestations of chirality in low-temperature NMR spectra. The consideration of the chirality of specific conformations can be especially important when discussing the mechanism of drug action, since the pharmacological activity can be due to one chiral conformation.

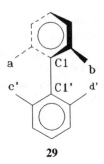

**29**

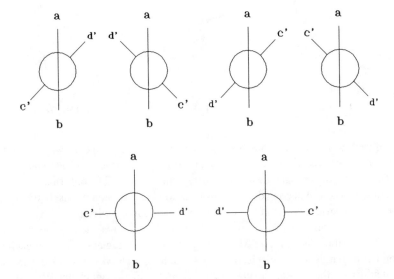

**Scheme 5.1**   The possible diastereomeric structures of **29**.

5.  The determination of the sense of chirality in certain molecules such as **157** according to the CIP classification is unclear, since three substituents on the central carbon atom C9b do not differ.

**157**

6. Among other purposes, the classification of chirality should serve to count the stereoisomers due to chirality. This task has not been recognized by the authors, and the CIP classification did not aim at it.

In addition, a misleading notation of the sense of the cyclophane chirality should be mentioned [27]. Namely, in the system **151** the sign of the $XC1C2C3$ angle is determined and denoted by R or S for the positive or negative sign, respectively. A subscript p for planar is sometimes added to distinguish this notation from the standard CIP one. This is a rather misleading notation since, for certain substituents $X$, $R_p$ corresponds to S chirality determined in the usual way by considering the priority of four $X$, C1, C2, and C3 atoms.

## 5.3 An Outline of the Modification of the CIP Classification with Some Examples

The preceding discussion shows that there was a necessity to modify the CIP classification. This task has been carried out in such a way that as little as possible has been changed in the CIP classification [23]. As mentioned before, the following discussion is based on the general condition that a molecule is treated as a rigid system of point masses connected by bonds. A discussion of some dynamic aspects of molecular chirality is indispensable, but the models of dynamically averaged structures will also be treated as the rigid ones. Such an approximation is sufficient for a discussion of chemical shifts in dynamic NMR spectra, but it clearly fails when effects associated with dipolar interactions are studied. Only systems consisting of atoms with coordination numbers less than or equal to four will be analyzed. Such a limitation does not represent any serious restriction for the isomer count, since for molecules possessing atoms with higher coordination numbers it is enabled by very effective mathematical methods based on graph theory. Moreover, factorization of molecular complexes with coordination numbers higher than four into elements of chirality is not unequivocal, and the designation of stereoisomers for such complexes is a serious problem.

A detailed proposal of a modification of the CIP classification was presented by Dodziuk and Mirowicz [23]. Here it will only be sketched briefly. A few examples illustrating the differences between two approaches, that is, the original CIP and that of DM, will be given.

The basis of the definitions of the elements of chirality forms the Möbius observation [28] that in a three-dimensional space there are only two different arrangements of four-point objects not lying in one plane. The first one presented in Fig. 5.3(a) corresponds to an asymmetric atom on which an irregular tetrahedron, called a simplex, can be built. Another one corresponding to three nonplanar consecutive vectors, usually bonds, representing a screw is shown in Fig. 5.3(b). It should be stressed that there is an essential difference between these arrangements. To determine the sense of chirality of the first one, a system of preference of substituents such as R/S proposed in the CIP classification for a center of chirality has to be

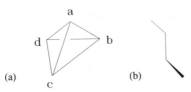

**Figure 5.3**  Two spatial arrangements of four noncoplanar points: (a) a simplex; (b) a screw.

introduced, while for the arrangement of the second kind the chirality sense P or M is defined by the sense of the screw, and no substituents at all should be involved [29]. However, these two arrangements bear a similarity since a tetrahedron can be built on both of them and the second one corresponds to the first one with connectedness constraints added. These similarities yield as a consequence the possibility of an equivocal description of chirality for certain molecules, such as allenes **147** and spiranes **149,** as having either centers or axes of chirality.

   The existence of only two nonplanar four-point arrangements in three-dimensional space indicates that only two different elements of chirality are necessary instead of three elements (a center, an axis, and a plane of chirality) and two additional types of chirality (conformational chirality and helicity) introduced in the CIP classification. The newly introduced elements of chirality are (1) a *center of chirality* analogous to that introduced in the CIP classification but differing from the CIP one in some rare implementations and (2) the second element, which, contrary to the axis of chirality of the CIP classification, has been called a *helical axis*. The axis is defined by three noncoplanar, usually consecutive, vectors forming a screw. As stated earlier, most molecules possessing asymmetric atoms are treated in the same way in the original CIP classification and the modification. In the latter the R/S system of preference of substituents is retained since it precisely describes configuration on an asymmetric atom and is deeply rooted in chemical practice.

   A few differences in both approaches are illustrated in the examples of **157** and **158.** The former molecule is classified as exhibiting a center of chirality by Cahn, Ingold, and Prelog (although the atoms proximate to their central atom are identical) and is thought to have helical axes in the proposed modification. For the conformations depicted in the formula only one axis is independent, and chirality descriptors can be found conveniently if one looks at its molecular model. Then the screw directions are clearly seen in the model, such as (H9b, C9b, C3a, C4)M, etc., for the depicted configuration of **157,** and we believe that this simplicity, instead of considering rather vague differences among the substituents on the central atom, is an argument in favor of this method of classification. Similarly, we believe that the determination of a sign of the screw formed by C1, C2, C3, C4 atoms using the molecular model of **158** proposed in the modification is simpler than the determination of the R/S chirality of four asymmetric atoms C1, C8, C9, C4. It should be stressed that the center of chirality can be ill-localized, and the R/S designation of its sense of chirality does not depend on the localization and sometimes even on the existence of the central carbon atom involved. For instance, the chirality of tetra-

(1S,4S,8R,9R) or (1,2,3,4)M

158

substituted adamantanes **98** and **150** is the most precisely described by one center of chirality situated somewhere within a tetrahedron formed by the four substituents.

Similarly to the R,S notation, the sense of a helical axis in a molecule should be given together with the numbering of atoms defining this axis or as a single P or M letter descriptor for screwlike or propellerlike molecules to denote overall chirality in agreement with the CIP recommendation. In the latter case (e.g., for hexa-helicene **76** or for [4.4.4]propellane **152**) an inspection of molecular models clearly reveals the corresponding screws if one looks at the models along the helix axis for helicenes or in the direction of $C_n$ symmetry axes for **152**, **157**, and **158**. It is obvious that not all conformations of *n*-butane **28** are chiral. The *trans* one **28a** is achiral, while the (+)- and (−)-*gauche*-butanes **28b** and **28c** are chiral. This mole-cule provides an example of conformational chirality, which is very important in many cases. It may be particularly important when the mechanism of action of a conformationally flexible chiral drug is analyzed. At room temperature **28** exists as a mixture of rapidly interconverting conformers. As discussed in detail in Chapter 2, due to the different time scales of the experimental techniques, the average structure is "seen" by NMR, while separate bands corresponding to *trans*- and *gauche*-**28** are detected in IR.

As stated earlier, within the CIP classification [9] for tetrasubstituted biphenyls **29**, only the structure averaged over internal rotation around the central $C_{sp^2}\!-\!C_{sp^2}$ bond was taken into account. Under this unformulated assumption the chirality of the parent biphenyl (and that of the spiro[5.5]undecane **57a,b**) has been overlooked. On the other hand, the chirality of **29** with frozen internal rotation (Scheme 5.1) cannot in general be classified using the Prelog and Helmchen definition, since it lacks the achiral skeleton precluding the application of their definition. In the proposed modification, depending on the experimental conditions, either the frozen conformations **29a–d** or the averaged ones **29e** and **29f** have to be considered. In the second, simpler case one can assign R or S chirality to the system on the basis of the priorities of the a, b, c′, and d′ substituents. Alternatively, the P or M symbols can be assigned to the system on the basis of the sign of a torsional angle. For instance, the symbol (c′,C1,C1′,a)M would describe **29e**. Two elements of chirality are necessary to describe the chirality of **29a–d** correctly. This could be both R (or S) symbols accompanied by the symbols of the substituents such as (a,b,c′,d′)R and the P (or M) one determined as above. Thus, a full description of the chirality sense of **29a** could be (a,b,c′,d′)R, (a,C1,C1′,d)M. For unsubstituted biphenyl **29** (a = b = c′ = d′) only the latter symbol has to be retained.

As said before, only two elements of chirality are sufficient to describe molecular chirality. Therefore, according to the modification there is no need to introduce the plane of chirality, helicity, and specific conformational chirality for **28, 29, 76, 159,** or nonplanar **160.** The sense of chirality of all these molecules can be described in terms of helical axes.

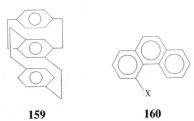

**159**                                      **160**

As described in Ref. 23, the modification of the CIP classification allows one to count stereoisomers due to the existence of the elements of chirality. However, the proposed formula is of importance only for very small and/or highly symmetrical molecules.

As mentioned earlier, the definitions of the modified elements of chirality are given in Ref. 23. Simplified rules for their choice seem to be of value to conclude this section. First, as in the original CIP classification, asymmetric carbon atoms are centers of chirality in the proposed modification. The R/S symbols are to be used to determine the sense of their chirality. Second, every four atoms a,b,c,d not lying in a plane can be used to define a helical axis. Then a screw formed by lines connecting these atoms determines a helical axis. The P/M symbols for the axis should be given together with the symbols of its determining atoms. The latter is necessary because of aboundance of such axes in molecules.

## 5.4  A Comparison of CIP Classification and Its Modification

The advantages of modifying the CIP classification presented here in comparison to the original version of the classification are as follows:

1. Instead of three (or five if helicity and conformational chirality are taken into account), only two elements of chirality (i.e., a center and a helical axis) are introduced here, and the elements are inherently related to the properties of a three-dimensional space.
2. The emphasis is placed on a precise description of the conformation of the molecule under investigation in accordance with the IUPAC rules on stereochemistry [30] but contrary to the CIP classification, in which definite conformations and averaged structures are confused during the presentation.
3. The presence of helical axes in some cyclic molecules such as **157, 158,** for which the assignment of the chirality elements had to be changed is easily seen in molecular models and conceptually seems to be much simpler than the idea of the presence of a center of chirality in these molecules.

4. A dichotomic use of the R/S and P/M systems of chirality descriptors proposed in the Prelog and Helmchen work has not always been followed in practice. Analogously, we propose to use the R/S symbols to describe senses of the centers of chirality and P/M to describe senses of chirality of helical axes. Such an approach allows one to overcome some discrepancies encountered with the use of the symbols in the original version of the CIP classification, but it cannot remove the duality of the description of chirality sense present in substituted allenes **147** and some other molecules discussed above.

5. The small extension of the R/S and P/M systems of chirality descriptors [23] (demanding naming of the atoms defining the element under consideration) allows for one simple and suitable for computer-codeable description of stereoisomers that are, due to the existence of the elements of chirality in a molecule under investigation.

6. A formula relating the number of independent chiral centers $n1$ and those of helical axes of different multiplicity $n2$, $n3$, etc., respectively, and the maximum number of stereoisomers due to the existence of the elements in the molecule is given, but its practical importance is limited to very small and/or highly symmetrical molecules, since in most of them there are many mutually interdependent elements of chirality, and the choice of independent ones is difficult and equivocal.

## References

1. E. H. Mann and J. L. Bada, *Annu. Rev. Nutr.*, **7** (1987) 209.

2a. A significant amount of some D-amino acids have been found not only in microorganisms, insects, and marine invertebrates [2b] but also in vertebrates. Recently, D-serine **138b** was found at a high D/L ratio in the brains of carp, frog, chick, mouse, rat, and bull [2b]. Its presence in human plasma was related to renal diseases [2c]. Moreover, the high specificity of the reactions in living organisms called for the existence of a special enzyme D-amino-acid oxidase.

2b. Y. Nagata, K. Horiike, and T. Maeda, *Brain Res.* **634** (1994) 291, and references cited therein.

2c. Y. Nagata, R. Masui, and T. Akino, *Experientia* **48** (1992) 986.

3. A. J. MacDermott, "The Electroweak Origin of Chirality," communication at the 5th International Conference on Circular Dichroism," Pingree Park, Colorado, August 18–22, 1993. Weak nuclear interactions do not conserve parity, and few calculations have been reported showing that they can be responsible for the prevalence of homochiral molecules in nature.

4. T. Ohmoto, K. Ikeda, and T. Chiba, *Chem. Pharm. Bull.* **30** (1982) 2780; B. J. Baker and R. G. Kerr, *Top. Curr. Chem.* **167** (1993) 1.

5. H. P. Abelson, *Researches in Geochemistry,* Vol. 2, Wiley, New York, 1967.

6. K. Bauer, D. Garbe, and H. Surbung, *Common Frangrances and Flavor Materials,* VCH Publishers, New York, 1990, p. 51.

7. B. Testa, *Chirality* **1** (1989) 7.

8. E. Fischer, *Ber.* **52A** (1919) 129.

9. R. S. Cahn, Ch. Ingold, and V. Prelog, *Angew. Chem.* **78** (1966) 413.

10. K. Mislow and P. Bickart, *Isr. J. Chem.* **15** (1976/77) 1.

11. K. Mislow and J. Siegel, *J. Am. Chem. Soc.* **106** (1984) 3319.

12a. The mothers who took this medicine during their pregnancy gave birth to babies with crippled extremities, for instance, with palms growing out of their arms.

12b. M. N. Cayen, *Chirality* **3** (1991) 94.

13. D. F. Arago, *Mem. cl. sci. math. phys. Inst. Imp. France* **12** (1811) 115.

14. J. B. Biot, *Bull. Soc. Philomath., Paris* (1815) 190.

15. L. Pasteur, Lectures delivered before the Societe Chimique de France, January 20 and February 3, 1860.

16. J. H. van't Hoff, *Nederl. Sci. Exactes Nat.* (1874) 445.

17. J. A. LeBel, *Bull. Soc. Chim. Fr.* **22** (1874) 337.

18. G. E. McCasland, "A New General System for the Naming of Stereoisomers," Chemical Abstracts, Columbus, OH, 1953.

19. V. Prelog and G. Helmchen, *Angew. Chem. Int. Ed. Engl.* **21** (1982) 567.

20. McCasland defined stereogenic elements by considering permutations of substituents on an atom. This definition was generalized by Mislow and Siegel [11] and Hirschmann and Hanson [21].

21. H. Hirschmann and K. Hanson, *Top. Stereochem.* **14** (1983) 183.

22. E. L. Eliel, *Stereochemistry of Carbon Compounds*, McGraw-Hill, New York, 1962, p. 29.

23. H. Dodziuk and M. Mirowicz, *Tetrahedr. Asym.* **1** (1990) 171.

24. D. M. Walba, *Tetrahedron* **41** (1985) 3136.

25. C.-O. Dietrich-Buchecker and J.-P. Sauvage, *Angew. Chem., Int. Ed. Engl.* **28** (1989) 189.

26. H. Dodziuk, *J. Chem. Soc., Perkin Trans.* 2 (1986) 249.

27. K. Schlögel, *Top. Curr. Chem.* **125** (1985) 29.

28. A. F. Möbius, "Der barycentrische Calcul," *Gesammelte Werke*, Vol. 1, Leipzig, 1885, cited in Ref. 19.

29. It should be stressed that the sense of the screw does not depend on the direction in which this screw is observed.

30. Commission on Nomenclature of Organic Chemistry. Rules for the Nomenclature of Organic Chemistry. Section E: Stereochemistry (Recommendations 1974).

# CHAPTER
# 6

# A Short Overview of Standard Structures of Organic Molecules

Eliel [1], Eliel and co-workers [2], and Mislow [3] published more than 20 years ago comprehensive monographs on the conformational analysis of standard organic molecules. Since then, some books on stereochemistry dealing mainly with typical organic molecules have appeared [4–6]. Therefore, the structural features of such molecules will only be briefly reviewed here, while those of novel, recently synthesized systems will be discussed in the next two chapters.

## 6.1. Hydrocarbons

### 6.1.1 Alkanes

Let us recall once more that the tetrahedral arrangement around a carbon atom forming the basis of the whole of organic stereochemistry is an idealization. As discussed in Chapters 2 and 4, even methane in its ground vibrational state does not possess strictly $T_d$ symmetry and exhibits a very small but measurable dipole moment. Nevertheless, in the following ideal molecular structures will be primarily discussed, since in most cases they are a good representation of real molecules.

The geometrical parameters for alkanes exhibit remarkable similarity. C—C bond lengths in standard molecules are usually equal to 152–154 pm, and those of C—H bonds lie within 104–110 pm [1–6]. Angles are equal to ca. 112°, 110°, and 105° for CCC, CCH, and HCH bond angles, respectively. As established by Pitzer et al. [7], internal rotation around the $C_{sp^3}$—$C_{sp^3}$ bond is hindered, and only certain values of torsional angles are allowed. This hindering is a purely quantum phenom-

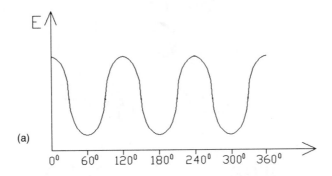

(a)

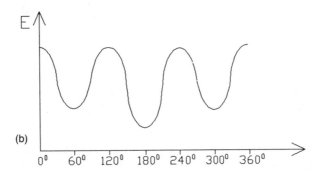

(b)

**Figure 6.1** The energy dependence on torsional angle for (a) ethane and (b) n-butane.

enon and has no classical analogy. For instance, in ethane **161** ($X = Y = Z = X' = Y' = Z' = H$) repulsions between the hydrogen atoms bonded to the different carbon atoms account only for ca. 10% of the rotational barrier. Thus the barrier hindering internal rotation cannot be rationalized in terms of the repulsion of neighboring, but not directly bonded, atoms. As said before, the barrier in ethane [see Fig. 6.1(a)] is equal to ca. 12 kJ/mol, with three equal minima for 60°, 180°, and 300°. The analogous minima in *n*-butane **28** corresponding to slightly different values of the CCCC torsional angle [presented in Fig. 6.1(b)] have different energies, reflecting a higher stability of the *trans* versus the *gauche* conformer. For a chain hydrocarbon formed by n carbon atoms, the number of torsional isomers is equal to $3^{n-1}$, but the number of the independent isomers is lower.

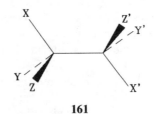

**161**

If one considers only possible values of CCCC torsional angles and does not take into account enantiomers, then three independent conformers are possible for *n*-pentane $CH_3(CH_2)_3CH_3$ (*trans,trans-*, *trans,gauche-*, and *gauche,gauche-*, since the *gauche,trans*-form is equivalent to *trans,gauche* and only changes the statistical weight of the latter). Similarly, there are five independent conformers of *n*-hexane, etc. The *gauche* conformers of *n*-butane (the "+" and "−" ones) are less stable than the *trans*, and in more complicated molecules the number of gauche arrangements is one of the main factors determining relative amount of conformers in the mixture. At room temperature in *n*-butane there is 59.1% of the *trans* and 40.9% of the *gauche* in the mixture [8]. By means of ED studies, Bartell and Kohl established that at room temperature there is 38.4% and 24.5% of the *all-trans* conformer in n-pentane and n-hexane, respectively [9]. It is interesting to note that for higher alkanes the amount of all-*trans* conformer further decreases with increasing chain length. It becomes less stable than the isomer with one *gauche* conformation at an end of the chain [10].

The barriers to internal rotation in alkyl-substituted ethanes are higher than the value for the parent compound. As can be seen from the following example, more distant substituents can exert a considerable effect on the barrier. For **162** the barrier is equal to 43 kJ/mol, while for **163** the corresponding value is equal to 58 kJ/mol [11]. Because of the thermal decomposition it was not possible to measure the barrier for 1,1,2,2-tetra-t-butylethane **164**. The *trans* rotamer of this molecule is of high energy and its *gauche* rotamers (i.e., enantiomers) were reported to be separable [12] similarly to the analogous tetraadamantyl compound **165** [13].

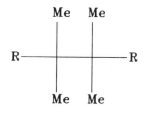

162 R=H   163 R=t−Bu

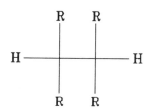

164 R=t−Bu   165 R=adamantyl

As mentioned before, alkanes usually exhibit remarkable constancy of their geometrical parameters. Voluminous substituents exhibiting strong nonbonded repulsions can change them. Due to these repulsions the CC bonds between methine and quaternary carbon atoms in tri-t-butylmethane are elongated to the abnormally high values of 161.1 pm [14]. According to the same study, the CCC bond angles within methyl groups are as small as 105°. It is interesting to note that the results of molecular mechanics calculations by Mislow indicate that, due to strong nonbonded repulsions, the symmetry of the still unknown tetra-t-butylmethane **131** should be lowered from $T_d$ to $T$ [15].

## 6.1.2 Cycloalkanes

Cyclopropane has an equilateral triangle structure ($D_{3h}$ symmetry) with an inter-atomic bond length of 151.0 pm and a corresponding bond angle of 60° [16]. According to Walsh's analysis of molecular orbitals [17], its bonds have partial double bond character.

As mentioned in Chapter 2, $\nu_{CD}$ stretching vibrations of monodeuterated cyclo-butane **47**-$d_1$ proved the nonplanar structure of the four-membered ring ($D_{2d}$ symmetry), with a CC bond length of 155.1 pm, a CCC bond angle of 88.5°, and a CCCC dihedral angle of 26 ± 3°. Ring inversion going through the planar transition state interchanges axial and equatorial substituents on the ring. The barrier to the process was found to be as low as 6 kJ/mol [18].

Two lowest-energy conformations of cyclopentane, the envelope ($C_s$ symmetry) **166** and half-chair **167** ($C_2$ symmetry), are very close in energy, and the barriers between them are lower than the energy of thermal movement $RT$ (2.5 kJ/mol at room temperature). Therefore, the molecule undergoes pseudorotation, visiting ten envelope and ten (five enantiomeric pairs) half-chair possible conformations. The planar form with an energy of about 20 kJ/mol is not a transition state in the pseudorotation process. Its CC bond length is equal to 154.6 pm, and the CCC bond angle amounts to 104.4° [19].

166                                                          167

Cyclohexane assumes mostly a rigid chair **22b** ($r_{CC}$ = 153.6 pm, $\vartheta_{CCC}$ = 111.4°, $\varphi$ = 54.9°) [20a] or flexible twist-boat **22c** conformations, with an energy difference of 23 kJ/mol between them. The free energy of activation of the intercon-version between them of 43.2 kJ/mol was found from ¹H NMR studies of cyclohex-ane-$d_{11}$ at a coalescence temperature of −61.4°C. By a complete line-shape analysis in variable temperature studies, $\Delta H^{\ddagger}$ and $\Delta S^{\ddagger}$ of 45.1 kJ/mol and 11.7 kJ/mol/de-gree, respectively, were found [20b]. At room temperature two possible chair forms interconvert rapidly, exchanging axial and equatorial substituents. The conformers with equatorial substituents are usually more stable, and the latter form prevails. The process of ring inversion in cyclohexanes is slow on the IR time scale, but it is rapid on the NMR time scale. Therefore, the bands characteristic of two conformers with axial and equatorial substituent positions are observed at room temperature by the first technique, while under the same conditions the averaged signals corre-sponding to the average structure are observed in NMR measurements [21]. The ring inversion can be frozen by lowering the temperature to ca. 233 K. Then the separate signals of the axial and equatorial protons can be detected by NMR mea-surements. The temperature dependence of the proton signals during the dynamic process of inversion of a saturated six-membered ring was shown in Chapter 2 for a

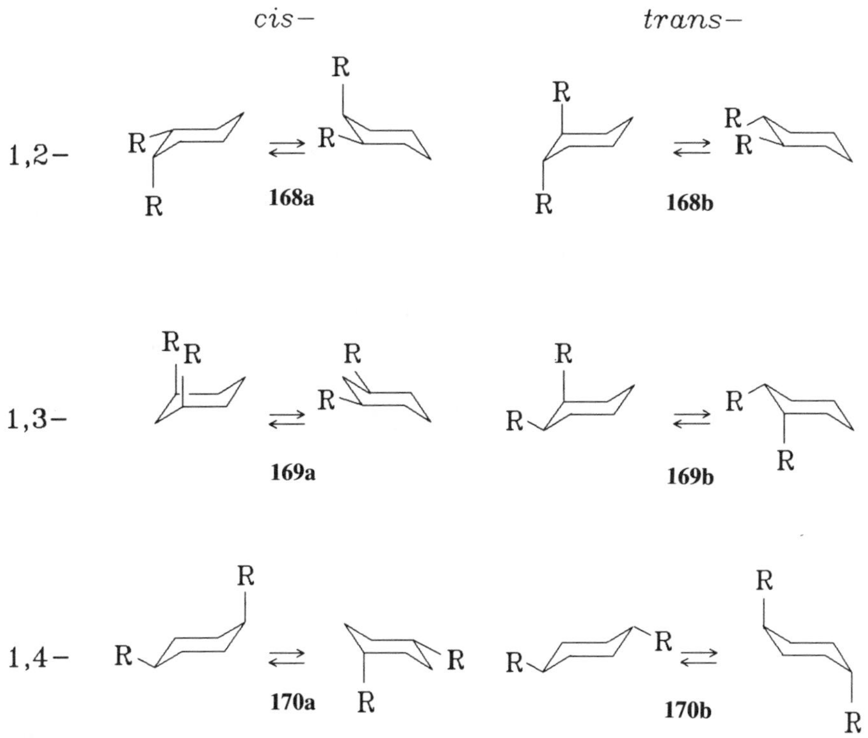

**Scheme 6.1**   Axial–equatorial equilibria for 1,2-, 1,3-, and 1,4-dialkylsubstituted cyclohexanes.

derivative of cyclohexane, tetraoxatrispirocompound **57c**. The conformational equilibria for 1,2-**168**, 1,3-**169**, and 1,4-**170**-substituted dialkylcyclohexanes are presented in Scheme 6.1. Due to steric repulsions, diequatorial conformers prevail for *cis*-1,3-, *trans*-1,2-, and *trans*-1,4-dialkylcyclohexanes [22].

Cycloheptane has a flexible structure with 76 ± 6% in the twist-chair form and 24 ± 6% in the chair form. $r_{CC}$ = 153.8 pm, $\vartheta_{CCC}$ = 114.8° [23]. Cyclooctane, cyclononane [24a], and cyclodecane have very flexible structures, with several minima close in energy separated by low barriers. The structure of the latter was determined by a joint application of ED and MM calculations [24b]. The average CC bond length in these molecules is equal to ca. 154 pm and the averaged bond angle values were as high as ca. 116.0° [24c]. Interesting dynamic phenomena are characteristic of larger rings, for which numerous conformations of comparable energy exist [24c].

## 6.1.3 Bicyclic Hydrocarbons

Bridgehead carbon atoms in bicyclobutane **171** are inverted; that is, their four substituents lie in one hemisphere [25]. Thus they do not conform to a tetrahedral arrangement of substituents around a tetravalent carbon atom. Such unusual systems

171                                              172

will be discussed in the next chapter. *Cis* substitution at the junction is more stable than the *trans* in small bicyclic hydrocarbons. An introduction of a 1,3 bond causes [3.1.0]bicyclohexane to assume a *cis* conformation **172** [26].

There has been an interesting discussion as to whether [2.2.2]bicyclooctane **173** has the highly symmetrical $D_{3h}$ or the skew $C_3$ structure. By means of MM calculations, Schleyer et al. showed that the former structure represents the minimum, which is very shallow, since the molecule undergoes large-amplitude torsional vibrations [27].

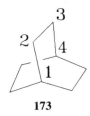

173

Bicyclo[4.4.0]decanes **25, 26,** better known as decalines, formally bear the name *perhydronaphthalenes*. They are constituent parts of many more complicated organic molecules. With the *trans* arrangement on the linking bond, there is a rigid *trans*-decalin structure **26**, while with the *cis* arrangement it is *cis*-decalin that undergoes rapid ring inversion [28]. The invertomers **25a** and **25b** of the latter are mirror images; therefore, they are enantiomers as well. Thus they form diastereomeric complexes with chiral β-cyclodextrin **14b**. The manifestation of chiral recognition of *cis*-decalin invertomers by the CyD in $^{13}$C NMR spectra at low temperatures was analyzed in Chapter 2 [29]. The in–out isomerism possible in much larger bicyclic systems will be discussed in Section 6.3.1.

Cyclopropane rings in spiropentane **118** ($D_{2d}$ symmetry) lie in perpendicular planes [30]. The average structures of higher spiranes are composed of two planar rings perpendicular to each other. Of course, in their frozen structures the rings are nonplanar. As discussed in Chapter 5, this fact has not been taken into account in the Cahn, Ingold, and Prelog classification of chirality, leading to the overlooking of the chirality of spiro[5.5]undecane **57a,b** and analogous systems [31]. Within this classification the chirality of such spiranes could only be induced by an appropriate substitution.

## 6.1.4 Alkenes

The main geometrical peculiarity of the nonovercrowded molecules having an isolated double bond is that it lies in a plane with its neighboring atoms. The barrier to

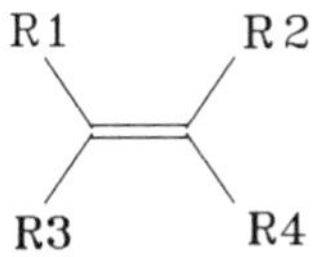

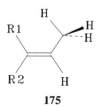

**174**

nonplanar distortions of ca. 240 kJ/mol in ethylene **174** ($R1 = \cdots = R4 = H$) is so large that most of its derivatives are planar. The C=C bond length in this molecule is equal to 133.9 pm, and it increases with substitution. The CCH bond angle in the same molecule is equal to 121.3° [32a]. An interesting comparison of the ethylene geometry established by various experimental techniques was reported by Cole and Gilson [32b]. The ideal 120° is reached in the planar $BF_3$ [33]. The changes in the angles around the $C_{sp^2}$ carbon atom are illustrated by the series **174b** ($R1 = Me, R2 = R3 = R4 = H$), 1,1-dimethylethylene **174c** ($R1 = R3 = Me, R2 = R4 = H$), and tetramethylethylene **174d** ($R1 = \cdots = R4 = Me$). The $HC_{sp^2}H$ bond angle is equal to 117.2°; due to nonbonding repulsions the corresponding value for the $C_{sp^3}C_{sp^2}C_{sp^3}$ bond angle decreases to 115.8° for **174c.** The so-called buttressing effect of the methyl substituents on the second carbon atom in **174d** diminishes this value to 112.2° [34].

Favored orientations of substituents in propenes **175** merit attention, since in spite of unfavorable nonbonded repulsions a $C_{sp^3}H$ bond is oriented *cis* to the double bond [35]. Such a preference influences the barriers to methyl group rotation in substituted propenes, resulting in much lower values of the barrier in *cis*-substituted propenes **175** ($R2 = H$) than the value for the parent compound [36]. In agreement with model considerations, the barriers in *trans*-propenes **175** ($R1 = H$) are close to those in propene itself.

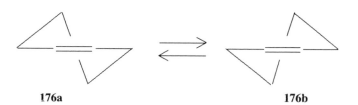

**175**

Cyclohexene **176** assumes a rapidly inverting half-chair conformation ($C_2$ symmetry). Recent studies have shown that the barrier to interconversion between two enantiomeric forms lies within a 35.3–50.8 kJ/mol range [37].

**176a**                                                    **176b**

### 6.1.5 Allenes and Alkynes

The carbon skeleton of allene **117** ($D_{2d}$ symmetry) is linear, but its two $C_{sp^2}C_{sp^2}H_2$ fragments lie in perpendicular planes. The $C_{sp^2}C_{sp^2}$ bond length of 131.3 pm is slightly smaller than the corresponding value for ethylene [38].

Alkynes are linear, and the triple $C_{sp}C_{sp}$ bond is the shortest (121.2 pm for acetylene **43** of $D_\infty$ symmetry [39]) among carbon–carbon bonds. The neighboring methyl group with a $C_{sp}C_{sp^3}$ bond length of 147.0 pm rotates freely [40]. As will be discussed in the next chapter, acetylenes incorporated into medium rings easily distort from linearity.

### 6.1.6 Conjugated Double Bonds

Similarly to the conjugated enones **45,** the most stable rotamer of butadiene is the planar *s-trans* one **177**. The amount of the second conformer in the mixture is small but could be enhanced by a sudden cooling of hot (700–1200 K) vapor of **177,178** to 30 K, allowing for UV and IR studies [41]. This small amount rendered experimental studies difficult, leading to long and vigorous discussions as to whether the second butadiene isomer has the planar *s-cis* **178** or the nonplanar *gauche* structure. To our knowledge, the discussion ended favoring the latter possibility [42]. The bonds in the *s-trans* conformer of butadiene exhibit a characteristic variation of their lengths. Its central bond has a length of 146.5 pm, intermediate between the values for ethane and ethylene, while the formally double bonds of 134.5 pm become only slightly longer in comparison to the ethylene $C_{sp^2}{=}C_{sp^2}$ bond. The $C_{sp^2}{-}C_{sp^2}{=}C_{sp^2}$ bond angle in **177** is equal to 123.3° [43]. By elongation of the system of conjugated bonds, the influence of the conjugation is stronger. Thus the difference between bond lengths of formal double and single bonds diminishes, but it never goes to zero [44].

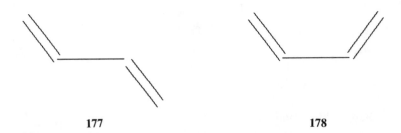

177                                                          178

The conjugation of a triple and a double bond and that of two triple bonds causes further shortening of the central bond (143.4 and 138.4 pm for **179** and **180,** respectively, in comparison to 146.5 pm in butadiene) [45].

179                                180

Cyclobutadiene **181** is unstable under normal conditions, but, as will be discussed in Chapter 10, it can be kept for days at room temperature in a "hemicarcerand flask," that is, as a supramolecular compound inside a macrocycle [46]. 1,3-Cyclohexadiene **182** is nonplanar [47], while its 1,4-unsaturated isomer **183** assumes either a planar or a considerably flattened boat form [48].

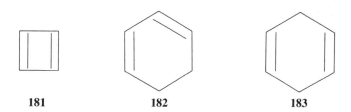

| **181** | **182** | **183** |

The nonrigidity of the structure of cyclooctatetraene **21** [49] was mentioned in Chapter 1. A nonplanar tub form was found on the basis of ED measurements. Two mechanisms of the interchange between these two forms, ring inversion and bond shift, are characterized by different activation energies.

## 6.1.7 Aromatic Systems

Probably everyone knows the story of how Kekule came upon the idea of the ring structure of benzene while dreaming about vigorously moving snakes that at a certain moment began to swallow their tails. Kekule proposed two such structures, **184a** and **184b.** Today we rather write the formula **184c,** indicating the average structure with all carbon atoms and all CC bonds equivalent. The $C_{sp^2}C_{sp^2}$ bond length in benzene is equal to 139.9 pm, and that of CH bond is 110.1 pm, while all bond angles in the molecule are equal to 120° [50]. The Dewar structures of benzene such as **184d** are much less stable than **184c.** Bond lengths in structure **184d** are given in the formula [51]. A superposition of a symmetry plane of the ring and the threefold axis of the methyl group results in a sixfold but very small (ca. 126 J/mol) barrier of internal rotation in toluene **61** ($X$ = H) [52]. The preferred value of the $C_{sp^2}C_{sp^2}C_{sp^3}C_{sp^3}$ angle in ethylbenzene **53** is ca. 70° [53].

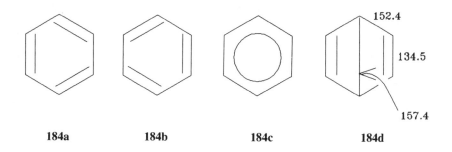

| **184a** | **184b** | **184c** | **184d** |

The biphenyl **35** [54] case was discussed in Chapters 2 and 5, where the differences in the angle formed by the rings between the solid-state planar structure and the gas-phase and solution structures with an angle of ca. 40° were discussed. The interactions of voluminous substituents in *ortho* positions in **29** remove this irregularity, leading to nonplanar structures both in the solid and gas phases [55]. The study of the reactivity of **29** [56] had a great impact on the understanding of nonbonded interactions and formulation of molecular mechanics. The torsional angle around the formally single $C_{sp^2}C_{sp^2}$ bond increases by further aromatic substitution. It increased to $46 \pm 5°$ in 1,3,5-triphenylbenzene **185** [57] and was found to be 90° in hexaphenylbenzene **186** [58].

185                                              186

The torsional angle values around the formally single bonds in *trans*-**104** and *cis*-**187** stilbenes, 43.2 and 32.5°, respectively, [59, 60] reflect stronger nonbonded repulsions in the former molecule.

187

## 6.1.8 Halogen-Substituted Hydrocarbons

The fluorine atom is considered to be only slightly larger than hydrogen. The chlorine atom is thought to be approximately as big as a methyl group, while bromine and iodine atoms are considerably larger [61]. However, the steric hindrance due to halogen substitution is not as big as one would expect judging from the van der Waals radii of the atoms. It is smaller because the length of the bond formed by the carbon and halogen atoms increases significantly in the series F, Cl, Br, I [60]. In addition, the electrostatic interaction between polar halogen atoms has to be taken into account. Due to the latter and to the steric hindrance, 1,2-dichloro-**161** ($X = X' = $ Cl; $Y = Y' = Z = Z' = $ H) and analogous 1,2-dibromoethane **161** ($X = X' = $ Br; $Y = Y' = Z = Z' = $ H) exist mainly in the *trans* conformation. The energy difference between the *trans* and *gauche* conformers is equal to 9.2 kJ/mol for the latter molecule [61]. It is smaller for the former (4.4 kJ/mol) [62].

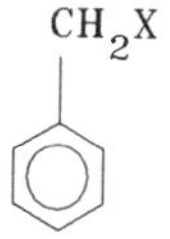

**188**
a: X=Cl
b: X=Br

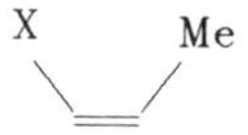

**189**

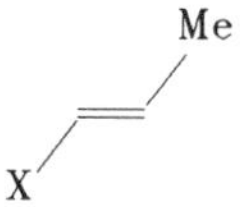

**190**

For $C_6H_5CH_2X$ **188** the twofold barrier to internal rotation around the $C_{sp^2}C_{sp^3}$ bond is equal to 8 kJ/mol [63].

Due to steric and electrostatic repulsion, the Z isomers of 1,2-dihalosubstituted ethylenes **189** are less stable than the E isomers **190** [64].

## 6.2 Compounds Containing Oxygen

### 6.2.1 Alcohols and Ethers

The configuration on oxygen atom resembles that on carbon with two substituents exchanged for lone pairs. Their smaller volume, together with the higher polarity of the oxygen atom, account for the difference between the configurations on these two centers. The values of CO bond lengths in various molecules are collected in Table 6.1 [65–68]. The OH bond length is 90–100 pm. The rotational barrier around the $C_{sp^3}O$ bond in methanol is considerably smaller than the corresponding barrier around the $C_{sp^3}C_{sp^3}$ bond (4.5 and 12 kJ/mol, respectively) [65].

Due to the presence of electron lone pairs, ethanol and higher alcohols exhibit *trans–gauche* isomerism about the $C_{sp^3}$—O bond analogous to that characteristic of $C_{sp^3}$—$C_{sp^3}$ bond [69]. The most stable conformation of dimethyl ether **191** is staggered, having $C_{2v}$ symmetry with the methyl groups tilted with respect to the corresponding CO bond [70].

In analogy with the stable conformations around the $C_{sp^2}C_{sp^3}$ bond in propene **175** derivatives, the most stable isomer of vinyl ether **192** [71a], 1-methoxycyclohexene **193** [71b], and anisole **194** [71c] is the planar *cis* one.

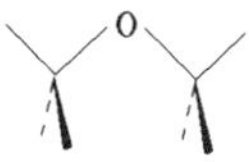

**191**

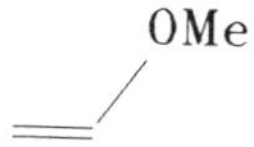

**192**

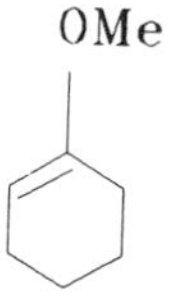

**193**

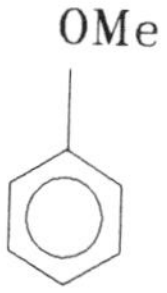

**194**

**Table 6.1**  CO Bond Lengths in Various Alcohols (in pm)

| R | 1 | R | 1 |
|---|---|---|---|
| $CH_3$ | 141–142 | $H_2C=C=CH$ | 137.5 |
| $H_2C=CH$ | 136.0 | $C_6H_5$ | 136.1 |
| $HC\equiv C$ | 131.3 | | |

### 6.2.1.1 The Anomeric Effect

Molecules having an electronegative substituent on the carbon atom in addition to oxygen exhibit a so-called *anomeric effect* consisting of a higher-than-expected stability of the *gauche* over the *trans* conformation with respect to the $C_{sp^3}$—O bond [72]. For instance, for chloromethoxy- and dimethoxymethane **195**, out of three possible conformations aa, ag$^{+/-}$, or g$^{+/-}$g$^{+/-}$, the enantiomeric g$^+$g$^+$ and g$^-$g$^-$ ones (**195a**) having $C_2$ symmetry are the most stable [72]. The effect plays a most important role for sugars for which an electronegative substituent (usually the oxygen atom) on a carbon atom next to the ring oxygen shifts axial–equatorial equilibrium toward a larger-than-expected amount of the axial substituent **196** [73]. By analyzing thermodynamic data for axial–equatorial equilibrium in dithiazines **197,** Booth and co-workers [74] found that the axial preference of the substituent Y can be determined by either enthalpic or entropic contributions.

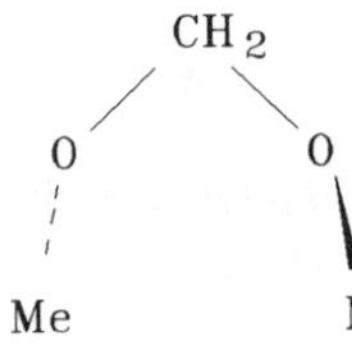

195a 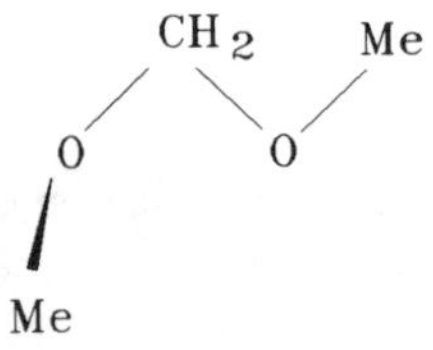            195b 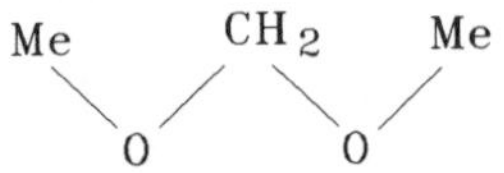            195c

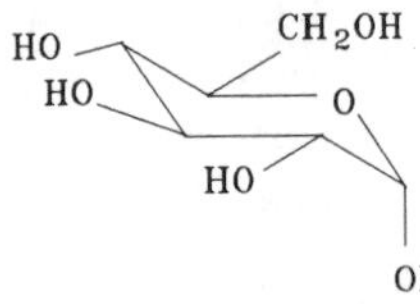

196 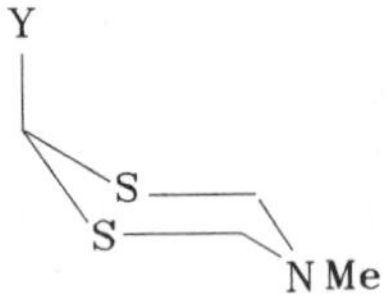            197

### 6.2.2 Aldehydes and Ketones

The length of the $C_{sp^2}$=O bond, 121 pm, is fairly independent of the neighboring substituents [75]. The conformations of alkyl groups in its vicinity are analogous to those in propene **175,** with the *cis* arrangement favored. For instance, for propan-1-al the *cis* conformer **198** is more stable than the *gauche* one **199,** with the barrier to interconversion between them equal to 3.8 ± 0.4 kJ/mol [76]. Similarly, for methyl ethyl ketone the conformer **200a** prevails with only a 5 ± 3% admixture of **200b** [77].

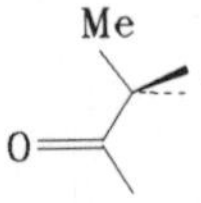

198 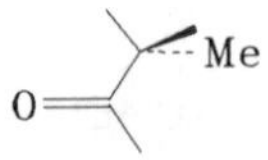            199 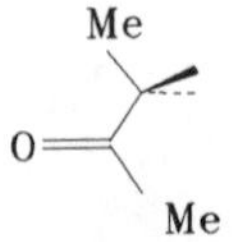            200a 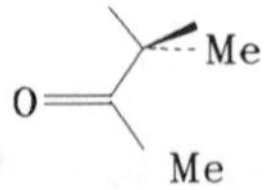            200b

The structure and conformation of acetylacetone is strongly influenced by a tautomerism involving hydrogen bond formation [78]. In the gas phase the enol *cis–s-cis* form **201a** is favored, while in solution the equilibrium between the enol and diketone **201b** forms strongly depends on the solvent polarity.

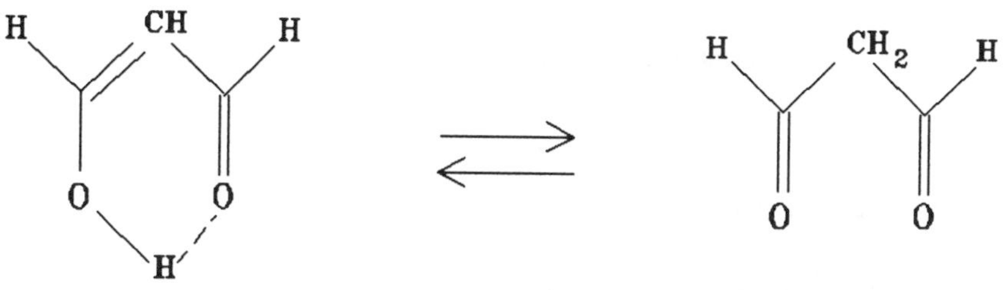

201a                                                    201b

*s-trans* and *s-cis* conformers are characteristic of conjugated enals and enones **44**. Large substituents in the 1 and 4 positions increase the amount of the latter in the mixture, as evidenced by their IR spectra discussed in Chapter 2 [83].

## 6.2.3 Carboxylic Acids and Their Esters

The carboxylic acids form very strong dimers **202** involving hydrogen bonds, even in the gas phase [79a]. Their IR spectra yield a very characteristic, easily recognizable pattern [79b]. At higher temperatures, by increasing pressure one can increase the amount of the monomer in the mixture [80]. The dimers are planar and have an inversion center. Their formation influences the geometrical parameters. For acetic acid $H_3COOH$ the $C={=}O$ bond length increases from 121.1 to 123.1 by dimer formation, while that of the C—C bond decreases from 136.4 to 133.4 pm [79]. In analogy to the propenes **175**, esters assume the *cis* conformation **203** [81].

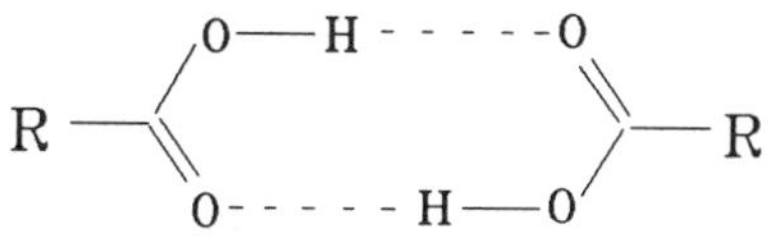

202

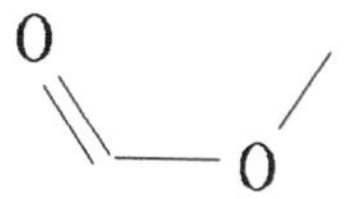

203

## 6.2.4 Peroxides

As mentioned in Chapter 4, hydrogen peroxide **73b** has a nonplanar structure with $C_2$ symmetry and reported values of the torsional HOOH angle of 119.1(1.8)° [82a] and 111.5° [82b].

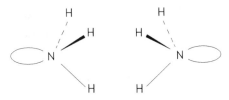

**Figure 6.2** Two pyramidal configurations of ammonia. The planar configuration is a transitional state.

## 6.3 Compounds Containing Nitrogen

### 6.3.1 Ammonia and Aliphatic Amines

The configuration on the nitrogen atom resembles that on carbon, with one hydrogen atom substituted by an electron lone pair. Thus ammonia $NH_3$ is nonplanar (Fig. 6.2). The rapid interconversion between its two nonplanar forms goes through a planar form. The enthalpy of its activation is equal to 25 kJ/mol [83]. One of the highest values of 98.7 $\pm$ 4.2 kJ/mol was found for the inversion barrier in oxaziridine **204** [84].

<br>

$$\begin{array}{c} O \\ H_3C \cdots \diagup \diagdown N \diagdown CH_3 \\ CH_2C_6H_5 \end{array}$$

**204**

<br>

The CN bond length in amines is equal to 145–147 pm. The CNC bond angle is close to 110°, and the methyl groups are tilted with respect to the CN bond [85]. The barrier hindering the rotation of the methyl group in amines is equal to ca. 8 kJ/mol [86].

In ethylenediamine **205** the *gauche* conformation with an NCCN torsional angle of 64 $\pm$ 1.6° prevails, representing 95% of the mixture [87]. An opposite trend is characteristic of hydroxylamines **206**, for which there is about 75% of the *trans* conformation in the gas phase [88].

<br>

$$H_2N \diagdown \diagup NH_2 \qquad\qquad \begin{array}{c} R1 \qquad\qquad R3 \\ \diagdown \qquad\qquad \diagup \\ N \text{——} O \\ \diagup \\ R2 \end{array}$$

**205**                                         **206**

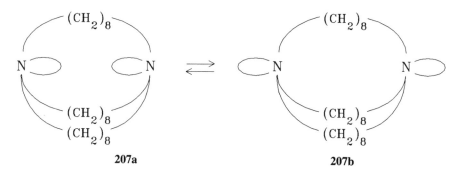

207a          207b

Bicyclic amines **207** exhibit the interesting phenomenon of "in–out isomerism" [89], which for larger cycles can also manifest itself for hydrocarbons [90]. It is interesting to note that the "in–out" isomer is much less stable than the "in–in" and "in–out" ones.

### 6.3.2 Aromatic Amines

Aniline **208** has a pyramidal nitrogen atom with an angle between the plane of the aromatic ring and that of the $NH_2$ group equal to 37.5° and an HNH angle of 113° [91]. A configuration close to planar on the nitrogen atom was found for dimethylaniline **209** [92] and triphenylamine **210** [93].

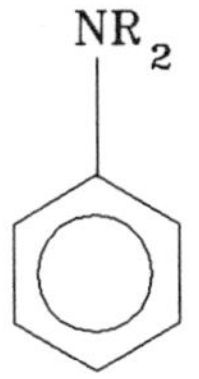

**208: R =H**
**209: R=Me**
**210: R=C₆H₅**

### 6.3.3 Hydrazines

The NN bond in $H_2NNH_2$ has a length of 144.9 pm, and its two NH bonds lie in perpendicular planes [94]. For 1,2-dimethylhydrazine ab initio MO calculations have been carried out to refine ED experimental data. Similarly to the bond arrangement in the parent hydrazine, the conformation with perpendicular CN bonds was found to be the most stable [$r_{NN} = 141.9(11)$ pm, $r_{CN} = 146.3(5)$ pm, $\varphi_{CNC} = 112(1)°$, $\tau_{CNNC} = 90°$] for dimethylhydrazine [95].

### 6.3.4 Amides of Carboxylic Acids

Amide **211** skeletons usually have a planar *s-trans* structure [96]. In tetramethylurea **212** the N-dimethyl groups are rotated out of the plane of the C=O bond by ca. 5° [97].

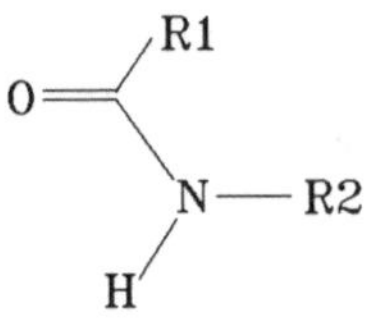

$$O = C \overset{R1}{\underset{\underset{H}{N - R2}}{}}$$

**211**

$$O = C \overset{\overset{Me}{N - Me}}{\underset{\underset{Me}{N - Me}}{}}$$

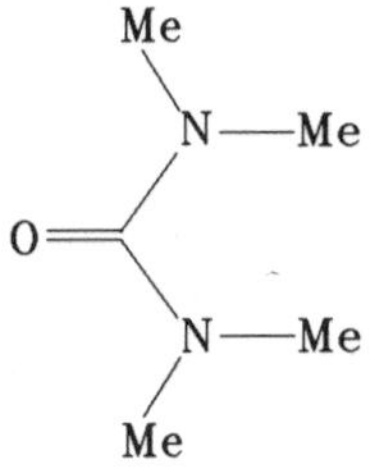

**212**

### 6.3.5 Imines, Azo Compounds, Nitryls, and Cumulenes Involving C=N Bonds

Imine $H_2C=NH$ **213** has a planar structure. $r_{CN} = 127$ pm; $\varphi_{XCN} = 110.5°$ [98]. In analogy to acetylenes **42**, nitryls **43** ($n = O$) are linear with a $C≡N$ bond length of 115–116 pm [99]. Isocyanates **214** and iso-thiocyanates **215** have nonlinear structures with a $C=NH$ bond angle of less than 140° [100].

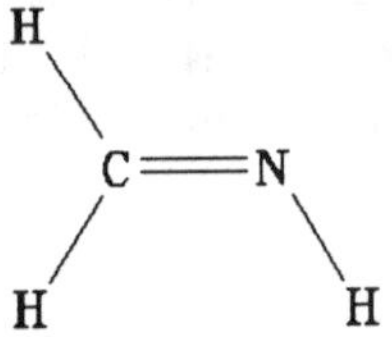

$$\overset{H}{\underset{H}{}}C = N \overset{}{\underset{H}{}}$$

**213**

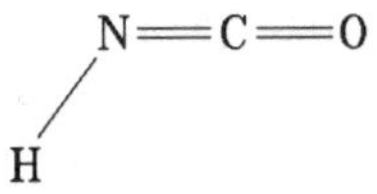

$$\underset{H}{}N = C = O$$

**214**

$$\underset{H}{}N = C = S$$

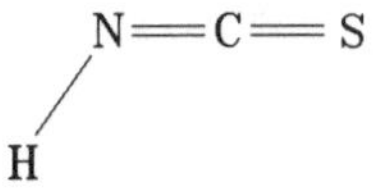

**215**

### 6.3.6 Nitro Compounds

The CN bond length in nitromethane is equal to 148.9 pm. The $NO_2$ group is planar. As is usual for sixfold barriers, the internal rotation in nitromethane is almost free (the barrier equals 25.2 J/mol) [101]. Even in $(CH_3)_3CNO_2$, $V_6 = 847.12$ J/mol [102]. Tetranitromethane $[C(NO_2)_4]$ was shown by ED and spectroscopic methods to have $S_4$ symmetry with the $NO_2$ planes inclined to the CN axis at 47.2° [103]. Nitroethylene is planar [104].

Nitrobenzene is planar [105], while the nitro groups in 1,2-dinitrobenzene are probably twisted [106].

## 6.4 Conclusions

On the basis of the studies discussed briefly in this chapter, the standard arrangements on the carbon atom can be briefly summarized in the following way:

1. Substituents on a tetravalent carbon atom are arranged tetrahedrally.
2. Double bonds and aromatic rings lie in one plane with their substituents.
3. Triple bonds are collinear with their substituents.

These observations form the foundations of classical conformational analysis. They are true for standard molecules but do not hold for many recently synthesized unusual molecules, which will be discussed in the next two chapters.

As mentioned at the beginning of this chapter, this review is intentionally by no means complete. An introduction of one or more heteroatoms or rings into an organic molecule or fusion with another ring, especially with a small one, causes a distortion of molecular geometry and influences the barriers involved. Original works or data on closely analogous molecules should be consulted for each particular system under study.

## References

1. E. Eliel, *Stereochemistry of Carbon Compounds,* McGraw-Hill, New York, 1962.

2. E. L. Eliel, N. L. Allinger, S. J. Angyal, and G. A. Morrison, *Conformational Analysis,* Interscience Publishers, New York, 1965.

3. K. Mislow, *Introduction to Stereochemistry,* Harper and Row, New York, 1973.

4. M. Nogradi, *Stereochemistry. Basic Concepts and Applications,* Akademiai Kiado, Budapest, 1981.

5a. E. Juaristi, *Introduction to Stereochemistry and Conformational Analysis,* J. Wiley and Sons, New York, 1991.

5b. D. Nasipuri, *Stereochemistry of Organic Compounds. Principles and Applications,* J. Wiley and Sons, New York, 1991.

6. E. L. Eliel and S. H. Willen, *Stereochemistry of Organic Compounds,* Wiley, New York, 1994.

7. J. D. Kemp and K. S. Pitzer, *J. Chem. Phys.* **4** (1936) 749; *J. Am. Chem. Soc.* **59** (1937) 276.

8. L. Rosenthal, J. F. Rabolt, and J. Hummel, *J. Phys. Chem.* **76** (1982) 817.

9. L. S. Bartell and D. A. Kohl, *J. Phys. Chem.* **39** (1963) 3097.

10. S. Mizushima, *Structures of Molecules and Internal Rotation,* Academic Press, New York, 1954.

11. J. E. Anderson and H. Pearson, *J. Am. Chem. Soc.* **97** (1975) 764.

12. S. Brownstein, J. Dunogues, D. Lindsay, and K. U. Ingold, *J. Am. Chem. Soc.* **99** (1977) 2073.

13. M. A. Flamm-ter Meer, H.-D. Beckhaus, K. Peters, H.-G. von Schnering, H. Fritz, and C. Rüschardt, *Chem. Ber.* **119** (1986) 1492.

14. H. B. Bürgi and L. S. Bartell, *J. Am. Chem. Soc.* **94** (1972) 5236.

15. L. D. Iroff and K. Mislow, *J. Am. Chem. Soc.* **100** (1978) 2121.

16. D. R. Lide, *J. Chem. Phys.* **33** (1960) 1514.

17. A. D. Walsh, *Trans. Farad. Soc.* **45** (1949) 179.

18. F. Takabayashi, H. Kambara, and K. Kuchitsu, Seventh Austin Symposium on Gas Phase Molecular Structure, Austin, Texas, 1978, p. 63.

19a. K. S. Pitzer, *J. Chem. Phys.* **5** (1937) 473; *ibid.* **8** (1940) 711; J. E. Kilpatrick, K. S. Pitzer, and R. Spitzer, *J. Am. Chem. Soc.* **69** (1947) 2483.

19b. B. Fuchs, *Top. Stereochem.* **10** (1978) 1.

20a. H. J. Geise, H. R. Buys, and F. C. Mijlhoff, *J. Mol. Struct.* **9** (1971) 447.

20b. F. A. L. Anet and A. J. R. Bourn, *J. Am. Chem. Soc.* **89** (1967) 760.

21a. M. Rey-Lafon and M. T. Forel, *J. Mol. Struct.* **29** (1975) 193.

21b. F. R. Jensen and C. H. Bushweller, *J. Am. Chem. Soc.* **88** (1966) 4279; F. R. Jensen, B. H. Beck, and C. H. Bushweller, *J. Am. Chem. Soc.* **91** (1969) 3223.

22. Ref. 2, Table 2.3.

23. F. A. L. Anet and J. S. Hartman, *J. Am. Chem. Soc.* **85** (1963) 1204; J. Laane, *Quart. Rev.* **25** (1971) 533.

24a. F. A. L. Anet and J. J. Wagner, *J. Am. Chem. Soc.* **93** (1971) 5266.

24b. R. L. Hilderbrand, J. D. Weiser, and L. K. Montgomery, *J. Am. Chem. Soc.* **95** (1973) 8598.

24c. J. Dale, *Top. Stereochem.* **9** (1976) 199.

25. M. D. Newton and J. M. Schulman, *J. Am. Chem. Soc.* **94** (1972) 767.

26. D. Seebach, *Angew. Chem.* **77** (1965) 119.

27. E. M. Engler, L. Chang, and P. von R. Schleyer, *Tetrahedron Lett.* (1972) 2525.

28. Ref. 5b, p. 306; R. Bucourt, *Top. Stereochem.* **8** (1974) 159. NMR measurements of the barrier to ring inversion in *cis*-decalin were reported by D. K. Dalling, D. M. Grant, and L. F. Johnson, *J. Am. Chem. Soc.* **93** (1971) 3678; B. E. Mann, *J. Magn. Reson.* **21** (1976) 17.

29. H. Dodziuk, J. Sitkowski, L. Stefaniak, J. Jurczak, and D. Sybilska, *J. Chem. Soc., Chem. Commun.* (1992) 207.

30. R. Boese, D. Bläser, K. Gomann and V. H. Brinker, *J. Am. Chem. Soc.* **111** (1989) 1501.

31. H. Dodziuk, *J. Chem. Soc. Perkin Trans.* 2 (1986) 249. See also the discussion of the CIP classification in the preceding chapter.

32a. E. Hirota, Y. Endo, S. Saito, K. Yoshida, I. Yamaguchi, and K. Machida, *J. Mol. Spectr.* **89** (1981) 223.

32b. K. C. Cole and D. F. R. Gilson, *J. Mol. Struct.* **28** (1975) 385.

33. G. Herzberg, *Infrared and Raman Spectra of Polyatomic Molecules,* van Nostrand, New York, 1946, p. 298.

34. L. V. Vilkov, V. S. Mastryukov, and N. I. Sadova, *Determination of the Geometrical Structure of Free Molecules,* Mir Publishers, Moscow, 1983, p. 84.

35. D. R. Lide, Jr., and D. Christensen, *J. Chem. Phys.* **35** (1961) 1374.

36. H. Dodziuk, *Bull. Acad. Sci. Polon., ser. chim.* **XXII** (1974) 75.

37. J. Laane and J. Choo, *J. Am. Chem. Soc.* **116** (1994) 3889.

38. F. Takabayashi, M. Sugie, and K. Kuchitsu, Seventh Austin Symposium on Gas Phase Molecular Structure, Austin, Texas, p. 103, 1978.

39. Y. Morino, K. Kuchitsu, and K. Fukuyama, *Acta Cryst.* **A25** (1969) 127.

40. C. C. Costain, *J. Chem. Phys.* **29** (1958) 864; A. Dubrulle, D. Boucher, J. Burie, and J. Demaison, *J. Mol. Spectr.* **72** (1978) 158.

41. M. E. Squillacote, R. S. Sheridan, O. L. Chapman, and F. A. L. Anet, *J. Am. Chem. Soc.* **101** (1979) 3657.

42. K. B. Wiberg and R. E. Rosenberg, *J. Am. Chem. Soc.* **112** (1990) 1509.

43. K. Kuchitsu, K. Fukuyama, and Y. Morino, *J. Mol. Struct.* **1** (1968) 463; A. R. H. Cole, G. M. Mohay, and G. A. Osborne, *Spectrochim. Acta* **A23** (1967) 909.

44. M. Traetteberg, *Acta Chem. Scand.* **22** (1968) 628.

45. M. Tanimoto, K. Kuchitsu, and Y. Morino, *Bull. Chem. Soc. Japan* **44** (1971) 386; A Rogstad, L. Benestad, and S. J. Cyvin, *J. Mol. Struct.* **23** (1974) 265; K. Fukuyama, K. Kuchitsu, and Y. Morino, *Bull. Chem. Soc. Japan* **42** (1969) 379.

46. D. J. Cram, M. E. Tanner, and R. Thomas, *Agnew Chem. Int. Ed. Engl.* **30** (1991) 1024.

47. M. Traetteberg, *Acta Chem. Scand.* **22** (1968) 2305.

48. S. Sabo and J. E. Boggs, *J. Mol. Struct.* **73** (1981) 137; H. Oberhammer, *J. Am. Chem. Soc.* **91** (1969) 10.

49. L. A. Paquette and T.-Z. Wang, *J. Am. Chem. Soc.* **110** (1988) 3663; F. A. L. Anet, J. J. R. Bourne, and Y. S. Lin, *J. Am. Chem. Soc.* **86** (1964) 3576.

50. K. Tamagawa, T. Iijima, and M. Kimura, *J. Mol. Struct.* **30** (1976) 243.

51. E. A. McNeill and F. R. Scholer, *J. Mol. Struct.* **31** (1976) 65; Y. Kobayashi and I. Kumadaki, *Top. Curr. Chem.* **123** (1984) 103.

52. K. C. Ingham and S. J. Strickler, *J. Chem. Phys.* **53** (1970) 4313.

53. P. Scharfenberg, B. Rozsondai, and I. Hargittai, *Z. Naturforsch.* **35a** (1980) 431.

54. O. Bastiansen, *Acta Chem. Scand.* **6** (1952) 205; C. P. Brock and K. L. Haller, *J. Phys. Chem.* **88** (1984) 3570; G. P. Charbonneau, and Y. Delugeard, *Acta Cryst., Sect. B* **32** (1976) 1420.

55. Ref. 5b, p. 80.

56. F. H. Westheimer, *J. Chem. Phys.* **15** (1947) 252.

57. Ref. 34, p. 107.

58. A. Almeningen, O. Bastiansen, and P. N. Skancke, *Acta Chem. Scand.* **12** (1958) 1215.

59a. M. Traetteberg, E. B. Frantsen, F. C. Mijlhoff, and A. Hoekstra, *J. Mol. Struct.* **26** (1975) 57.

59b. M. Traetteberg and E. B. Frantsen, *J. Mol. Struct.* **26** (1975) 69.

60. Ref. 5b, p. 225.

61. K. Tanabe, J. Hiroshi, and T. Tamura, *J. Mol. Struct.* **33** (1976) 19; A. Yokozeki and S. H. Bauer, *Top Curr. Chem.* **53** (1975) 104.

62. K. Kveseth, *Acta Chem. Scand.* **A28** (1974) 482.

63. L. V. Vilkov, Author's Abstract of Doctoral Dissertation, Moscow, MGU Press, 1968.

64. Ref. 34, p. 147; J. A. van Shaik, F. C. Mijlhoff, G. Renes, and H. J. Geise, *J. Mol. Struct.* **21** (1974) 17.

65. M. C. L. Gerry, R. M. Lees, and G. Winnewisser, *J. Mol. Spectr.* **61** (1976) 231.

66. S. Samdal and H. M. Seip, *J. Mol. Spectr.* **28** (1975) 193.

67. Ref. 34, p. 148.

68. D. G. Lister and P. Palmieri, *J. Mol. Struct.* **32** (1976) 355.

69. R. K. Kraker and P. J. Seibt, *J. Chem. Phys.* **57** (1972) 4060.

70. W. Benkis, P. H. Kasai, and R. J. Meyers, *J. Chem. Phys.* **38** (1963) 2753.

71a. P. Cahill, P. Gold, and N. L. Owen, *J. Phys. Chem.* **48** (1968) 1620.

71b. L. Schäfer, *Appl. Spectr.* **30** (1976) 123.

71c. H. M. Seip and R. Seip, *Acta Chem. Scand.* **27** (1973) 4024.

72a  Graczyk i Mikolajczyk, *Top. Stereochem.* **21** (1994) 159.

72b.  P. Deslongchamps, *Stereoelectronic Effects in Organic Chemistry,* Pergamon Press, Oxford, 1983.

72c.  R. U. Lemieux and N. J. Chue, *Abstr. Papers Amer. Chem. Soc. Meetings* **133** (1958) 31N.

73.  C. Romers, C. Altona, H. R. Buys, and E. Havinga, *Top. Stereochem.* **4** (1969) 39.

74.  E. Juaristi, E. A. Gonzales, B. M. Pinto, B. D. Johnson, and R. Nagelkerke, *J. Am. Chem. Soc.* **111** (1989) 6745.

75.  R. B. Lawrence and M. W. P. Stranberg, *Phys. Rev.* **83** (1951) 363; R. W. Kilb, C. C. Lim, and E. B. Wilson, *J. Chem. Phys.* **26** (1957) 1695.

76.  S. S. Butcher and E. B. Wilson, Jr., *J. Chem. Phys.* **40** (1964) 1671.

77.  M. Abe, K. Kuchitsu, and T. Shimanouchi, *J. Mol. Struct.* **4** (1969) 245.

78.  A. H. Lowrey, C. George, P. D'Antonio, and J. Karle, *J. Am. Chem. Soc.* **93** (1971) 6399.

79a.  R. V. Davies, A. G. Robiette, M. C. L. Gerry, E. Bjarnov, and G. Winnewisser, *J. Mol. Spectr.* **81** (1980) 93; B. P. van Eijck, van Opheusden, M. M. M. van Schaik, and E. van Zoeren, *ibid.* **86** (1981) 465; A. Almenningen, O. Bastiansen, and T. Motzfeldt, *Acta Chem. Scand.* **23** (1969) 2848.

79b.  K. Nakanishi, *Infrared Absorption Spectroscopy. Practical,* Holden-Day, San Francisco, 1972, pp. 43, 175–179.

80.  Ref. 34, p. 161.

81.  S. Cradock and D. W. H. Rankin, *J. Mol. Struct.* **69** (1980) 145.

82a.  G. A. Khachkuruzov and I. N. Przhevalskii, *Opt. Spectrosc. (USSR)* **46** (1979) 1034.

82b.  R. H. Hunt, R. A. Leacock, C. W. Peters, and K. T. Hecht, *J. Chem. Phys.* **42** (1965) 1931.

83.  A. Rand, L. C. Allen, and K. Mislow, *Angew. Chem., Int. Ed. Engl.* **9** (1970) 400; J. B. Lambert, *Top. Stereochem.* **6** (1971) 19.

84.  A. Mannschreck, J. Linns, and W. Seitz, *Ann.* **727** (1969) 224.

85.  H. K. Higginbotham and L. S. Bartell, *J. Chem. Phys.* **42** (1965) 1131.

86.  E. B. Wilson, *Advances in Chemical Physics,* Interscience, New York, 1959, Vol. 2, p. 370.

87.  *Molecular Structure by Diffraction Methods,* G. A. Sim and L. E. Sutton, Eds., London, The Chemical Society, Vol. 1, 1973, p. 59.

88.  S. Tsunekawa, *J. Phys. Soc. Japan* **33** (1972) 167; D. W. H. Rankin, M. R. Todd, F. G. Riddell, and E. S. Turner, *J. Mol. Struct.* **71** (1981) 171.

89.  C. H. Park and H. E. Simmons, *J. Am. Chem. Soc.* **90** (1968) 2431.

90.  C. H. Park and H. E. Simmons, *J. Am. Chem. Soc.* **94** (1972) 7184.

91.  Ref. 93, p. 59; D. G. Lister, J. K. Tyler, J. H. Hog, and N. W. Larsen, *J. Mol. Struct.* **23** (1974) 253.

92.  L. V. Vilkov and T. P. Timasheva, *Dokl. AN SSSR* **161** (1965) 351.

93.  Y. Sasaki, K. Kimura, and M. Kubo, *J. Chem. Phys.* **31** (1959) 477.

94.  Y. Morino, T. Iijima, and M. Murata, *Bull. Chem. Soc. Japan* **33** (1960) 46.

95.  W. J. Beamer, *J. Am. Chem. Soc.* **70** (1948) 2979; K. Kohata, T. Fukuyama, and K. Kuchitsu, *Chem. Lett.* **257** (1979).

96.  M. Kitano and K. Kuchitsu, *Bull. Chem. Soc. Japan* **47** (1974) 67; *ibid.* **46** (1973) 3048.

97. L. Fernholt, S. Samdal, and R. Seip, *J. Mol. Struct.* **72** (1981) 217.

98. R. Pearson, Jr., and F. J. Lovas, *J. Chem. Phys.* **66** (1977) 4149.

99. K. Karakida, T. Fukuyama, and K. Kuchitsu, *Bull. Chem. Soc. Japan* **47** (1974) 299.

100. Ref. 34, p. 183.

101. A. P. Cox and S. Waring, *J. Chem. Soc. Faraday Trans.* 2 **68** (1972) 1060.

102. P. R. R. Langridge-Smith, R. Stevens, and A. P. Cox, *J. Chem. Soc. Faraday Trans.* 2 **76** (1980) 330.

103. N. I. Sadova, N. I. Popik, and L. V. Vilkov, *J. Mol. Struct.* **31** (1976) 399.

104. P. Nösberger, A. Bauder, and Hs. H. Günthard, *Chem. Phys.* **1** (1973) 426.

105. J. Hog, L. Nygaard, and G. O. Sorensen, *J. Mol. Struct.* **1** (1970) 11; J. Trotter, *Acta Cryst.* **12** (1959) 884.

106. Ref. 34, p. 188.

# Pot-pourri of Saturated Pathologies

## 7.1 Hydrocarbons with Unusual Spatial Structures—Are They a Mere Curiosity?

In 1978 Liebman and Greenberg published a book on strained organic molecules [1a]. The chapter of this book dealing with unusual systems bore the title "Potpourri of Pathologies." Since then the number of "pathologies" has grown immensely, and syntheses of molecules with unusual structures constitute one of the most rapidly developing domains of organic chemistry. Numerous saturated and unsaturated hydrocarbons with nonstandard spatial structure, topological molecules mimicking knot, Möbius strip, nonrigid molecules, and mesoionic compounds have been synthesized recently. Theoretical and experimental studies on many other systems allow one to expect that the syntheses of many other atypical molecules will be achieved in the near future. In this chapter some nonstandard saturated hydrocarbons will be presented, while other unusual molecules will be discussed in the next chapter. Supramolecular complexes, which are also atypical structures from the point of view of classical organic chemistry, will be discussed in Chapter 10. Inevitably, the content of this and the next chapter partly overlaps with Vögtle's book, *Fascinating Molecules in Organic Chemistry* [1b]. However, the choice of molecules and the approach adopted in discussing them differ significantly from that of Vögtle.

The question "Why are molecules with unusual structure worth studying?" has to be addressed first. Everyone admits that these molecules are aesthetically pleasing and that their nonstandard syntheses present a challenge. At present, they do not have industrial applications, but this will undoubtedly change in the future [2]. (The short fullerene history provides examples of high expectations and unfulfilled promises of applications, which will be briefly discussed in Section 8.5.) In addition to their highly unusual reactivity, molecules with nonstandard spatial structure are of considerable cognitive significance since they allow a better understanding of the nature of the chemical bond.

The bond is a very useful intuitive concept, but it is difficult to apply to unusual systems such as nonrigid molecules, mesoionic compounds, supramolecular complexes (such as the complexes of cyclodextrins with hydrocarbons discussed in Chapter 10), and van der Waals complexes. As will be shown, the theoretical and experimental studies of propellanes, especially those of [1.1.1]propellane **5**, enabled a better understanding of the nature of chemical bonds. The question as to how far a bond can be distorted without being broken is fascinating. In turn, it evokes a second question as to the properties of such a distorted bond. These questions are not easy to answer since (1) quantum chemistry correctly describes only a molecule as a whole, and until very recently it was difficult or even impossible to describe precisely its fragments within a quantum model. (2) Calculations for large and unusual molecules require laborious ab initio computations involving configuration interactions and big basis sets with full geometry optimization, which are time consuming and beyond the reach of most research groups in spite of the rapid development of computers. Nevertheless, theoretical calculations play an ever-increasing role in the search for new unusual molecules. With the rapid development of computers today it is possible not only to rationalize experimental results but also to predict reliably the properties of systems that are hypothetical or difficult to study. This enables one to propose plausible synthetic targets on the basis of theoretical calculation. Sometimes such predictions are premature (as was in the case for fullerene $C_{60}$ **11** discussed in Section 8.5). When timely carried out, calculations can stimulate synthetic chemists, leading to an exciting development, as illustrated by Wiberg's propellane studies to be presented in the following section. Wiberg's successful quest for molecules having inverted carbon atoms [3] shows how a close interaction between theoretical and synthetic studies can invigorate the development of synthetic chemistry. Even when the calculations provide conflicting results, as were the estimations of the energy required for methane planarization discussed in Section 7.2.2, they can stimulate the syntheses (as happened in the case of fenestranes **63** [4] and paddlanes **216** [5]).

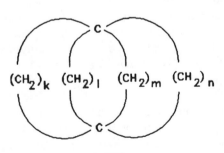

**216**

## 7.2 Saturated Hydrocarbons with Strongly Distorted Bonds [6]

### 7.2.1 Molecules Possessing Inverted Carbon Atoms

An inverted atom is, according to the Wiberg et al. definition [7], an atom having its four substituents in one hemisphere. Dodziuk [8a] has pointed out that this definition does not differentiate between inverted and pyramidal atoms (as in **5** and **235,** respectively). To make the distinction, inverted atoms should be defined as those having bond configurations obtained from the tetrahedral one by reversing the direction of one bond. Eight molecules having inverted carbon atoms, **5, 171, 217–222,** that is, bicyclobutane and small-ring propellanes having the general formula **64,** have been obtained [3]. [2.2.2]Propellane **223** represents in this respect a limiting case, since its bridgehead atoms lie approximately in a plane with its corresponding three neighbors. The properties of propellanes have been summarized in Ref. 3. Here, only their most important structural features will be presented, together with a fascinating story of the successful interaction between theoretical and synthetic chemistry in the propellane domain. It should be stressed that the ease of formation, stability, structure, vibrational and photoelectron spectra, and enthalpy of formation of [1.1.1]propellane **5** were all predicted by Wiberg and co-workers [10, 11] prior to its synthesis [3]. The correctness of the predictions followed soon. The strength of the central bond formed by two inverted carbon atoms in the molecule was vigorously debated; even its existence was questioned on the basis of quantum calculations [12]. This bond manifests its peculiarity in a derivative of [3.1.1]propellane **218** [13] and in a bicyclobutane derivative tricyclo[2.1.0.0²,⁵]pentane **224** [14] by the lack of so-called deformation density [15] in X-ray diffractograms.

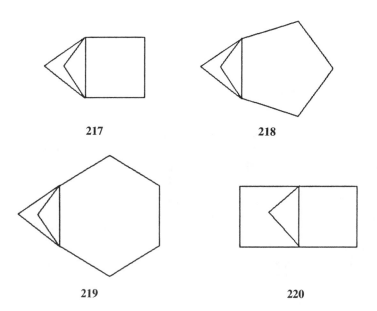

217                                         218

219                                         220

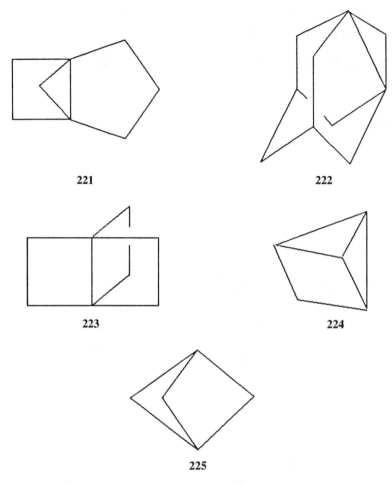

221

222

223

224

225

It is interesting to note that in spite of the highly unusual bonding pattern, the length (152.5 ± 0.2 pm) of bridge–bridgehead CC bonds differs from that in cyclopropane by only ca. 2 pm. More understandably, the length of the central bridgehead–bridgehead bond (159.6 ± 0.05 pm) is considerably longer [16]. The evidence that this bond really exists was provided by Feller and Davidson's calculations [17] of the bridgehead CH bond dissociation energy in bicyclopentane **225**, since the dissociation of the first bond requires more energy than that of the second one, 444 and 197 kJ/mol, respectively. The approach developed by Wiberg, Bader, and Lau [18] has shown the buildup of the charge between the bridgehead carbon atoms, thus proving once more the existence of the central bond in small-ring propellanes. The discussions on the existence and properties of the central bond in [1.1.1]propellane pointed out the inherent limitations of quantum chemistry, which yields results for a whole molecule but does not allow one to "extract" properties of molecular fragments from the calculations. As mentioned in Chapter 3, Bader and co-workers [19] recently developed a procedure to remedy this deficiency.

Except for bicyclobutane **171** and octabisvalene **68** (which can be considered as the bicyclobutane dimer), known molecules having inverted carbon atoms are all small-ring propellanes in which the inverted atoms are always present in symmetrical pairs. On the basis of model considerations and molecular mechanics calculations, Dodziuk has shown [20] that (1) [$k$.1.1]propellanes with $k > 4$ form an infinite family of propellanes having inverted carbon atoms. (2) Lower analogs of the Paquette [2.2.2]- and [3.2.1]geminanes **226–228**, respectively, [21] such as [1.1.1]geminane **229**, should have inverted carbons and exhibit no excessive strain. Tricycloheptane **230** should have only one such atom, while unsymmetrical geminanes **231, 232** should have two differing inverted atoms. On the basis of the MM calculations, the molecules have been shown to be relatively unstrained; thus they seem to be prospective targets. Their synthesis would considerably enlarge the rather exclusive group of molecules having inverted carbon atoms.

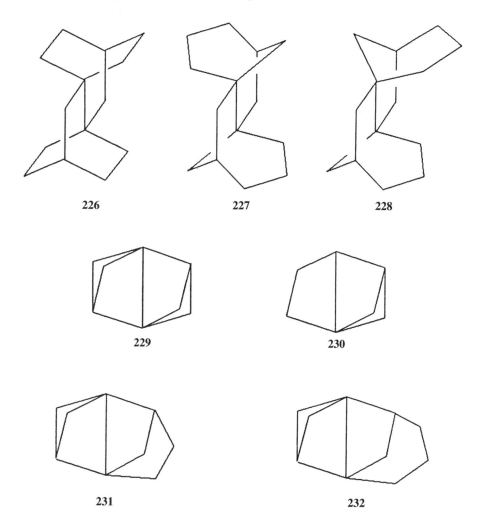

226          227          228

229          230

231          232

As discussed in Chapter 3, the possibility of a diagonal bond in cubane yielding **65** having two inverted carbon atoms was recently explored on the basis of ab initio quantum calculations [22] as well as that of the synthesis of highly strained face-fused dicubane **66** [23].

### 7.2.2 "The Planar Methane Problem" and the Pyramidal Carbon Atom

The problem of the synthesis of a molecule that has a carbon atom lying in a plane with its four substituents was formulated by R. Hoffmann et al. [24]. Although calculations on methane varying from those based on extended Hückel theory to ab initio ones (discussed in Refs. 6 and 8) and geometry optimization indicated that a pyramidal arrangement of bonds around a carbon atom was more stable than the planar one, most theoretical and experimental efforts in this field were focused on molecules with a planar configuration.

Although the estimates of the energy needed to enforce methane planarity varied from 400 to 1000 kJ/mol, many interesting possible synthetic targets have been proposed on the basis of theoretical considerations. Some of them were later synthesized, but none possessed either a planar or a pyramidal configuration on the carbon atom. For many years the theoretical and synthetic efforts have been concentrated on small-ring $[k.l.m.n]$fenestranes **63** $(k,l = 4, 5)$ [4] and $[k.l.m.n]$paddlanes **216** [5]. At first [4.4.4.4]fenestrane **63a** (also known as windowpane) was thought to possess a planar configuration at the central carbon atom. As often happens, this erroneous idea triggered a lot of studies on small-ring fenestranes. The ab initio calculations by the Schulman group [25] showed that, out of two possible configurations on the central carbon atom, the pyramidal **233** is less stable than the quasitetrahedral **234**. Despite massive synthetic efforts aimed at the synthesis of [4.4.4.4]fenestrane **63a** described in Ref. 4, [5.4.4.4]fenestrane **63** $(k = 5, l = m = n = 4)$ [26] seems to be the smallest synthesized molecule in this series.

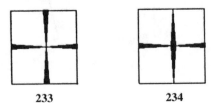

233                                    234

On the basis of molecular mechanics calculations, bowlane **235** [8a] was proposed as a plausible synthetic target that should have a pyramidal arrangement on the central carbon atom. The MM method yielded a symmetrical $C_{4v}$ structure as a minimum, while by both semiempirical AM1 [8b] and ab initio QC [9a] calculations this structure was shown to be a saddle point. According to the latter calculations, the minimum structure has lower, $C_{2v}$ symmetry. However, this molecule seems to be relatively unstrained, with the configuration on the central carbon atom close to planar. Interestingly, MM calculations for its analogue **236** [8b] exhibited an arrangement very close to planar. Taking into account its high strain, the latter molecule is believed to be very difficult, if possible, to synthesize. However, on the basis of his QC calculations, Radom came to the opposite conclusion [9b].

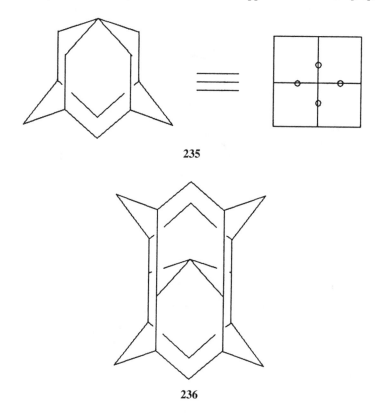

**235**

**236**

## 7.2.3 Molecules with a Planar Cyclohexane Ring

It took about half a century to prove and accept the premature Sachse hypothesis of the chair conformation of cyclohexane published about a hundred years ago [27]. Then it became one of the foundations of classical stereochemistry and was turned into a kind of dogma. Therefore, the planar structure of the saturated rings in molecules **237–239** and that in some complexes of **240** ($X$ = N, O) [28] has not gained due recognition. *trans*-Tris-σ-homobenzene **241** was synthesized, and the same planar structure was proposed for its central ring [29]. No sound argument was given in favor of this statement, but the close-to-planar structure of its hexa-methylderivative **242** [30] seemed to validate this claim. The same structure was proposed for six-membered rings in hypothetical hexaasterane **243** [31], hexa-prismane **244** [32, 33], and truncated tetrahedrane **245** [33] on the basis of molecular mechanics calculations. The latter molecule was also studied by means of ab

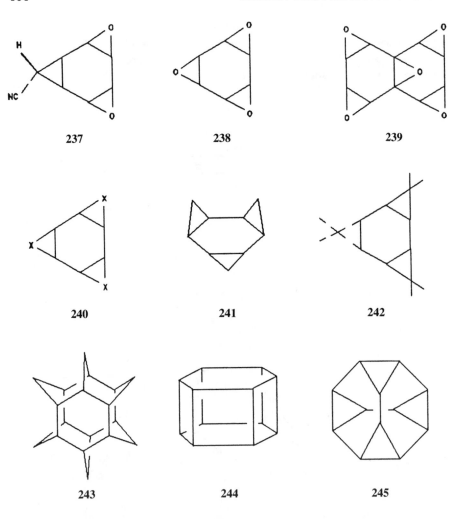

initio quantum calculations [34]. Diademane **246** was synthesized [35] but its geometry has not been determined. Both MNDO [36] and ab initio [37] calculations for this molecule yielded a planar structure for the six-membered ring. The molecules **244–246** belong to the $C_{2n}H_{2n}$ family of saturated cage compounds, and as such they will be discussed in the next section.

Many molecules with the distorted chair conformation of six-membered rings are known [38]. A *cis*-fusion with a smaller ring was proposed as the driving force for ring flattening [39] on the basis of a systematic molecular mechanics study of bicyclic molecules **247–249**, *cis*- and *trans*-bicyclooctanes **250, 251**, respectively, coronanes **6, 252, 253,** and the coronanes lacking one ring **254, 256, 257.** Highly strained hypothetical molecules **253, 257** have been included in the series for the sake of completeness. At the time of the study, the molecules **247–251, 255,** and a

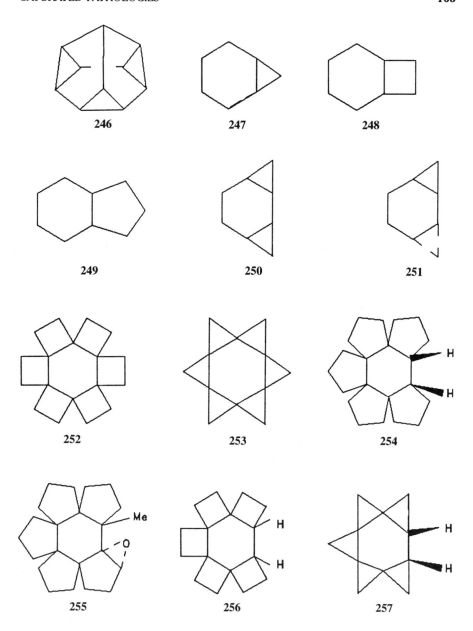

246          247          248

249          250          251

252          253          254

255          256          257

derivative of **256** were known (see Ref. 42 for the pertinent references), while [6.5]coronane **6** was synthesized later by the Fitjer group [40]. The sum of absolute values of the torsion angles within the six-membered ring, $\alpha = \Sigma \, \omega$, was chosen in Ref. 39 as a measure of ring planarization. Very few experimental data were available to compare with the results of the calculations. However, the calculated results can be summarized as follows: (1) In cases when a comparison with experimental data is possible, the calculated values of α agree with the experimental ones. (2) The agreement is much worse for the calculated values of bond lengths. (3) The latter combined with the condition for ring closure led to the poor accuracy of the calculated bond angles. It should be stressed that the planarization of the central ring in [6.5]coronane **6** was correctly predicted [39] prior to its synthesis [40]. It is surprising that the MM2 parametrization developed for hydrocarbons with standard geometry correctly describes the distortions of six-membered ring toward planarity. The results of the calculations discussed in the next section seem to indicate that MM2 also correctly reproduces the distortions of cyclohexane ring towards larger puckering. Thus molecules possessing a cyclohexane ring fused with one or more smaller rings are also reliably described by the MM method introduced in Section 3.3.

### 7.2.4 Saturated Cage Compounds $C_{2n}H_{2n}$

One could say in jest that studies of saturated cage compounds $C_{2n}H_{2n}$ started in ancient Greece since carbon skeletons of the molecules **133, 7,** and **8** represent the ideal polyhedra tetrahedron, cube, and dodecahedron discussed by Plato and Pythagoras.

The molecules $C_{2n}H_{2n}$ consist of carbon atoms having three carbon and one hydrogen neighbor atoms each. In spite of this limiting condition, the number of isomers grows rapidly with increasing $n$. Only one member of the $C_{2n}H_{2n}$ family could exist for $n = 2$, **133,** and $n = 3$, **258;** there are three isomers for $n = 4$, cubane **7,** cuneane **67,** and octabisvalene **68,** 9 isomers for $n = 5$, 32 for $n = 6$, etc. [41]. Only the following molecules or their derivatives are known: **133** [42], **258** [43], **7** [44, 2], **67** [45], **68** [46], pentaprismane **259** [47], diademane **246** [48], octahedrane **71** [49], and dodecahedrane **8** [50]. The field is rapidly developing, encompassing the synthesis of new exciting molecules, their subsequent physicochemical studies, as well as theoretical investigations, which often precede the syntheses. The failures are usually not described in the literature, but the unsuccessful syntheses of hexaprismane **244** and truncated tetrahedrane **245** have been reported [51]. Within the group of saturated cage compounds $C_{2n}H_{2n}$ a smaller group of [$n$]prismanes can be considered consisting of triprismane **258,** cubane **7** being tetraprismane, pentaprismane **259,** hexaprismane **244,** heptaprismane **260** [52], etc. This group of compounds will be discussed at the end of this subsection.

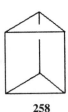

**258**

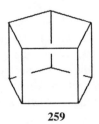

**259**

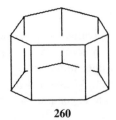

**260**

## 7.2.4.1 The $C_4H_4$ Group: Tetrahedrane 133

The highly strained structure of tricyclo[1.1.0.0$^{2,4}$]butane has attracted chemists for over 70 years [53]. Its structure and the possibility of its synthesis have been intensively explored theoretically. The molecule was found too strained to be synthesized. Only two derivatives of tetrahedrane, tetra-t-butyl **4** [54a] and tetra-t-butyl(trimethylsilyl)tetrahedrane [54b], seem to have been synthesized. The former, unexpectedly stable one has been extensively studied. An X-ray study [55] has shown that the distances between the carbon atoms in the tetrahedrane core of **133**, 149.7 pm, are very close to the corresponding value in cyclopropane. Both $^1$H and $^{13}$C NMR spectra of the molecule [56] reflect the high symmetry and peculiarity of the structure. Its structure and unusual stability have been the subject of comprehensive studies. Using semiempirical quantum calculations, Minkin et al. [57] discussed the synthesis, reactions, and stability of both **4** and **133**, explaining the high kinetic stability of the former as being due to the unfavorable steric repulsions in the product of its rearrangement tetra-t-butyl-cyclobutadiene **261**. On the basis of ab initio RMP2 calculations using a 6-31G* basis set with full geometry optimization, Schulman et al. have shown [58] that for medium-sized hydrocarbons second-order correlation energies yield reasonable values for the heats of formation. Indeed, their calculated value for cubane **7** is very close to the experimental one. However, because of the lack of appropriate thermochemical data, their calculated values for the most interesting molecule under study, that is, for tetrahedrane, could not be checked against experimental value.

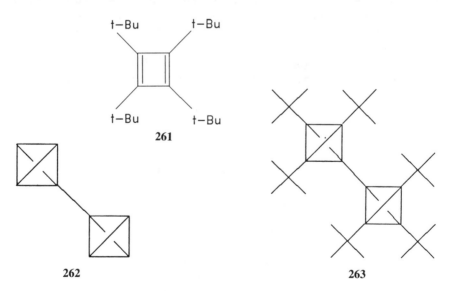

Hypothetical coupled bitetrahedranes **262** and **263,** as well as known coupled bibicyclobutane **264,** bicubanes **265,** and coupled bicyclopentane **266,** were shown to exhibit a remarkably short central CC bond either theoretically or on the basis of X-ray studies [59].

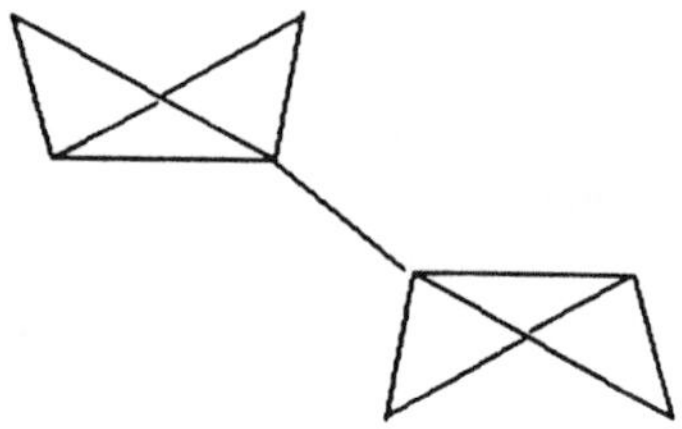

264

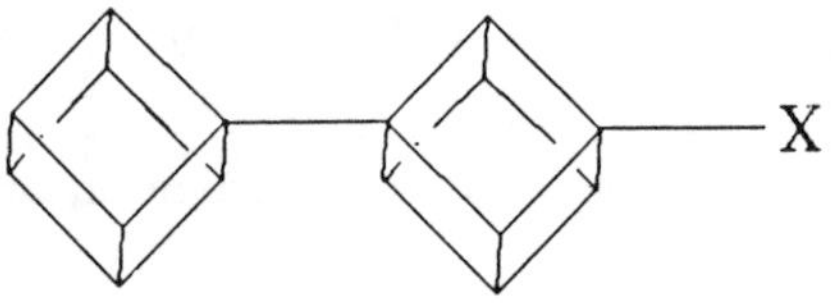

265

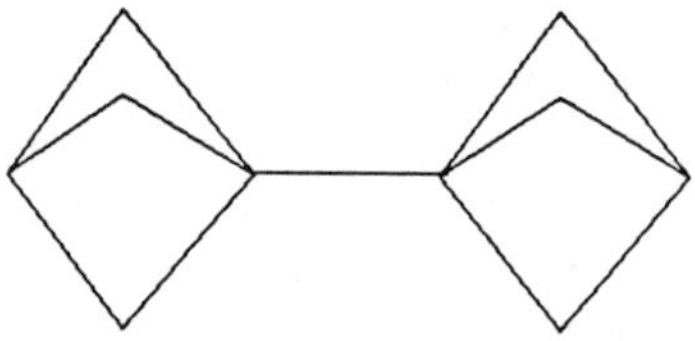

a: X=H
b: X=Br

266

## 7.2.4.2 The $C_6H_6$ Group: Triprismane

The molecule **258**, also called Ladenburg benzene or [3]prismane, has been known since 1973 [43], but has been studied very little experimentally. Its geometry and HOF have not been determined. Only the structure of its hexamethyl derivative was investigated [60]. Numerous theoretical studies [57, 61, 62] showed that high barriers have to be overcome to execute the transition to thermodynamically more stable isomers of this and higher prismanes.

A synthesis of **267** en route to propella[$n_3$]prismane **268** was reported by Gleiter and Treptow [63]. Thus new molecules with a triprismane core are expected to be synthesized soon.

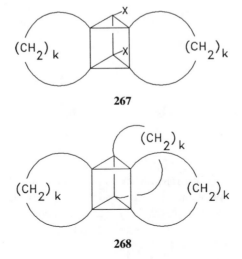

267

268

### 7.2.4.3 The $C_8H_8$ Group: Cubane 7, Cuneane 67, and Octabisvalene 68

Because of its high symmetry and early synthesis [44], cubane has attracted great interest on the part of both theoreticians and experimentalists. Therefore, this molecule is probably the most studied among those discussed in this chapter. The synthesis and chemistry of cubane as well as its physicochemical properties and theoretical studies have been recently reviewed by Griffin and Marchand [64]. There is only one kind of CC bond in the molecule, and this has enabled accurate studies of its length. Unfortunately, the results obtained by means of X-ray analysis [65], electron diffraction [66], and microwave spectra of cubane-$d_1$ [67] differed considerably, yielding 155.1(3), 157.5(1), and 157.08 pm, respectively.

In view of the low accuracy of determinations of H atom positions, the differences for CH bond lengths were even larger. To resolve these discrepancies and to determine a quadratic force field for the molecule, Hedberg et al. [68] combined an ED study of 7 in the gas phase at 344 K with microwave rotational constants for cubane-$d_1$. The discussion in the latter study clearly illustrates the difficulties encountered in a comparison of experimental results obtained by different modern experimental and theoretical methods discussed in Chapter 2 of this book and in Ref. [69]. The use of combined ED–MW data and their sophisticated analysis allowed the authors to estimate the equilibrium value $r_e$(CC) as 156.18(40) pm and that of $r_e$(CH) as 109.60(1.30) pm. A comparison of equilibrium bond lengths (with estimated effects of molecular vibrations) with the calculated values revealed that both MINDO/3 [70] and ab initio investigations using the 6-31G* and the STO-3G basis sets [71] showed good agreement with the experimentally derived value of $r_e$(CC). Somewhat surprisingly, MM calculations parametrized for standard hydrocarbons yielded reasonable values for the bond lengths and excellent agreement with the experimental value of the heat of formation (expt. 623 kJ/mol [72]; calc. 624 kJ/mol [73]).

The interesting MM study of the products of cubane hydrogenolysis was mentioned in Chapter 3. Similarly, the course of the reaction of a $[\pi^2 + \pi^2]$ photochemical ring closure of the *syn*-tricyclo[4.2.0.0$^{2,5}$]octa-3,7-diene derivative **269** yielding propella[3$_4$]prismane **270** [74] was elucidated in a combined AM1 and MM study [75]. A comparison of this reaction with the failure to obtain cubane from tricyclooctadiene **271** in the same reaction showed that the latter one involves a prohibitively large increase of strain and that the ordering of the frontier molecular

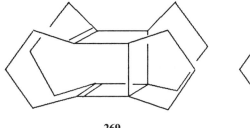

269

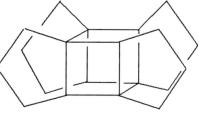

270

orbitals governed by the double bond–double bond distance is the most important factor in this reaction. Potential energy calculations by MM2 indicated that the trimethylene bridges in **270** are so flexible that they should appear flat on the NMR time scale. Indeed, $^{13}$C NMR spectra [75] exhibited only three signals compatible either with $\mathbf{D_{4h}}$, $\mathbf{C_{4h}}$, and/or $\mathbf{D_{2h}}$ structures or an equilibrium mixture of low-energy conformers that averages to these symmetries.

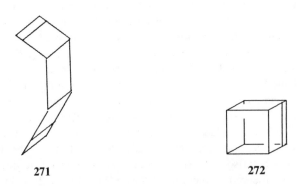

271                                                272

Some of the studies on cubane derivatives are the subject of other sections. An abnormally short central bond length found in X-ray study [76] of a 4-bromoderivative of bicubane **265b** [77] was mentioned in Section 7.2.4.1. The possibility of the existence of cubane with a diagonal bond **65** and that of face-fused dicubane **66** was discussed earlier in Chapter 3, while the synthesis of cubene **272** will be mentioned in the next chapter.

Although cuneane **67** [45] and a derivative octabisvalene **68** derivatives [46] are known, only a few experimental data for the molecules have been published. They have been summarized in reviews [6] together with those on unsaturated members of the $C_8H_8$ family [78]. Except for one semiempirical quantum study, all theoretical calculations discussed in the latter article yielded cubane (which is [4]prismane) as the most strained member of the family, but there is no agreement concerning the relative stability of **67** and **68**. The lack of experimental thermochemical studies and ab initio calculations for these molecules calls for more studies. In some papers the existence of inverted carbon atoms discussed in Section 7.2.1 (the bridgehead ones in bicyclobutane units) in octabisvalene **68** has been overlooked.

### 7.2.4.4 The $C_{10}H_{10}$ Group: Pentaprismane 259 and Diademane 246

As mentioned before, within this group **259** [47] and **246** [48] are the only known molecules. Their experimental studies are very scarce. Only a photoelectron spectrum of pentaprismane was reported [79]. By means of semiempirical quantum calculations, Minkin [57] and Spanget-Larsen and Gleiter [36] predicted **246** to have planar cyclohexane rings. The same conclusion has been reached by Schulman et al. on the basis of ab initio calculations [40]. Therefore, in spite of a lack of experimental confirmation of its spatial structure, it is sure that **246** belongs to the group of molecules discussed in Section 7.2.3.

### 7.2.4.5 The $C_{12}H_{12}$ Saturated Cage Molecules: Hexaprismane 244, Truncated Tetrahedrane 245, and Octahedrane 71

In spite of extensive attempts [50, 51], until recently the only known molecule belonging to this group was a methyl derivative of heptacyclo[6.4.0.0²,⁴.0³,⁷.0⁵,¹².0⁶,¹⁰.0⁹,¹¹]dodecane **71** [80]. As mentioned in Section 3.3, on the basis of the MM calculations, the parent compound was proposed to be a prospective synthetic target in 1987 [33]. Its synthesis was reported in 1993 by the de Meijere group [81a]. The structure of **71** determined by X-ray analysis [81b] confirmed the unusually high puckering of the central cyclohexane ring predicted by the MM calculations. The torsional angle ω of 75.2° computed for the central six-membered ring in the molecule was much larger than the standard value of 56° established for the chair conformation of cyclohexane. Surprisingly, the agreement of the calculated value with the experimental one was within a few degrees. Secohexaprismane **273** was obtained by Mehta and Padma [82]. However, an attempt to synthesize **244** as a part of a larger system **274** proved unsuccessful [51b].

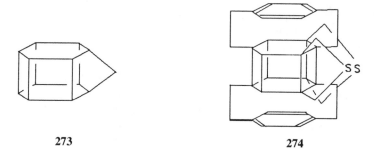

273                                    274

The most comprehensive study of the $C_{12}H_{12}$ family was carried out using MM by Dodziuk and Nowinski [33]. The calculations were performed for only a few isomers of this group. As discussed earlier, out of those studied **71** was found to be the most stable and was synthesized soon thereafter [81]. Both MM [33, 83] and QC [84] calculations predict **D$_{6h}$** symmetry with planar cyclohexane bases for hexaprismane **244,** and according to all calculations its HOF should be much larger than that of the former molecule: QC, 641.4 kJ/mol; MM2, 541.3 kJ/mol. Therefore, the failures to synthesize it [51] are not surprising. The situation with truncated tetrahedrane **245** is not that clear. Its HOF values calculated using ab initio [85] and MM2 [33] methods are very close, 381 versus 364 kJ/mol, respectively. The latter value is only ca. 42 kJ/mol larger than that of **71**. In addition, the calculated vibrational frequencies [85] of truncated tetrahedrane **245** indicate the kinetic stability of this highly symmetrical molecule. Thus the unsuccessful attempts to synthesize it [51] cannot be explained at present. Schriver and Gerson [84] came to the same conclusions concerning **71, 244,** and **245** on the basis of ab initio calculations for a series of $C_{12}H_{12}$ saturated and unsaturated compounds. Although the MM and QC calculations [33, 84] were not carried out for all $C_{12}H_{12}$ isomers of saturated cage compounds, they indicate that hexaprismane **244** is one of the least stable molecules within the series.

### 7.2.4.6 [n]Prismanes

A few existing studies on higher prismanes will be mentioned in this section, in addition to the information on lower prismanes presented in former sections. Photochemical reactions leading to heptaprismane **260** analogues have been investigated experimentally and by MM2 calculations by Mehta et al. [52b], but the molecule remains unknown. A $D_{4d}$ structure has been proposed for octaprismane **275** by Reddy and Jemmis [83], while Jemmis et al. [86] and Miller and Schulman [87], on the basis of semiempirical AM1 and MNDO calculations, argued that this structure is not a local minimum. Miller and Schulman [87] calculated geometries and vibrational frequencies of [n]prismanes under the constraint of $D_{nh}$ symmetry, using basis sets varying from STO-3G to 6-31G* together with an RMP2 computation. The calculations yielded pentaprismane **259** as the most stable molecule within the series, followed by triprismane **258** and cubane **7.** After that, the HOF values increased considerably with increasing $n$. Thus the hypothesis by Dodziuk [73b] that, for a given $n$, [n]prismanes are one of the least stable molecules within the $C_{2n}H_{2n}$ family can be justified. With this respect, two jocular structures deserve mentioning. Following a witty idea of Eschenmoser, helvetane **276** and israelane **277** were proposed by Ginsburg in the April 1 issue of the *Nouveau Journal de Chimie* [88]. Notwithstanding, some theoretical groups have studied them [87, 89]. They even came to the conclusion that helvetane, being only slightly less stable than [12]prismane, may be capable of existence. However, we believe that QC calculations [84, 85] and MM studies [33, 73b] suggest that higher prismanes are not promising synthetic targets.

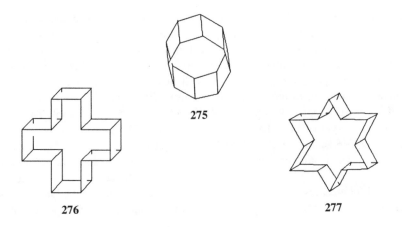

275

276　　　　277

## 7.2.4.7 The $C_{20}H_{20}$ Family: Dodecahedrane 8 and Pagodane 278

As said before, to our knowledge no members of $C_{2n}H_{2n}$ family with $n = 7-9$ are known, but Paquette [50, 91] and Prinzbach [90] succeeded in obtaining two molecules with $n = 10$, dodecahedrane 8 and pagodane 278, respectively. Together with fullerene 11, the former one belongs to the highest point symmetry group of icosahedron $I_h$, having 120 symmetry elements. As pointed out in Chapter 4, not long ago molecules having such a high symmetry were thought to be improbable of synthesis. In spite of its ball-like shape, 8 was found to form regular crystals with a distinct favored orientation [92]. This is different from the crystals of $C_{60}$ discussed in Chapter 2 and Section 8.5. Due to the high symmetry of 8, all its carbon atoms are equivalent. Interestingly, the dodecahedron cavity is too small to include even the smallest ion or atom.

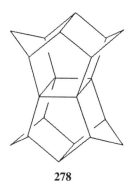

278

The obtaining of dodecahedron 8 [50] was one of the most laborious and intellectually demanding tasks in organic synthesis, which took almost 20 years. It provided us with a better understanding of the chemistry of polycyclic compounds, which are capable of undergoing the unusual rearrangements. The second known $C_{20}H_{20}$ molecule, pagodane 278, was originally thought to be a useful dodecahedrane precursor. Indeed, its functionalization allowed numerous efficient syntheses of specific dodecahedrane derivatives [93]. Unsaturated derivatives of 8 and 278 will be briefly discussed in the next chapter.

## 7.2.4.8 Some Other Nonstandard Saturated Hydrocarbons

The realm of nonstandard saturated hydrocarbons is unlimited, and it cannot be, even briefly, covered in one chapter. Therefore, only a few additional examples will be mentioned here.

Centropolyquinane is unknown, but Kuck and Schuster [94] reported the synthesis and properties of its unsaturated derivative, centrohexaindane 134. The latter molecules are very interesting from the point of view of graph theory, since they are topologically nonplanar; that is, their graphs cannot be drawn without an intersection of lines. The highly symmetrical known compound, propellacubane **125** [74], was mentioned earlier. As discussed in Section 3.3, [3]rotane **69** is an interesting example, since the lack of agreement between MM on one hand and an X-ray analysis on the other hand forced a repetition of experiments that confirmed theoretical results [95]. Some other rotanes **279–281** and mixed rotane **282** deserve mentioning here [96], as well as coronanes **89** [97], **6** [40], and the still-hypothetical **252.** The syntheses and extensive X-ray and dynamic NMR studies of rotanes and some of their analogues allowed Fitjer and co-workers to formulate conditions under which the central six-membered ring assumes the conformation of a flattened chair with a barrier to ring inversion as high as 136.0 kJ/mol [96]. With differing substituents as in **282,** the central ring assumes a twist-boat conformation [98].

Di-**283,** tri-**284,** and tetraasteranes **285,** as well as ditetraasterane **286,** are known

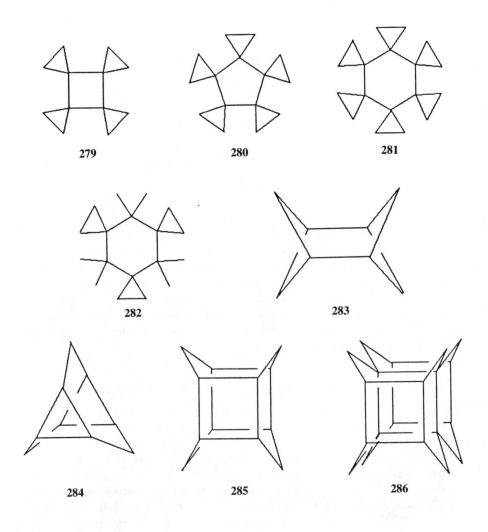

279

280

281

282

283

284

285

286

[99]. The symmetrically substituted tetraasteranes have a $C_4$ symmetry axis. As discussed in Chapter 4, fourfold symmetry axes are quite rare in organic molecules, since the cyclobutane ring is usually nonplanar. The hypothetical molecule hexaasterane **243** [31] was also discussed in Section 7.2.3 on planar cyclohexane rings.

An example of an alkylsubstituted cyclohexane will complete this chapter, showing that even ordinary molecules may not fit into oversimplified schemes. The preference for an equatorial orientation of a substituent is well established in monosubstituted cyclohexanes. However, all-*trans*-1,2,3,4,5,6-hexa-isopropylcyclohexane **91,** having an $S_6$ improper symmetry axis, was shown to exist in crystals in the chair conformation with all substituents in axial positions [100].

## References

1a. J. F. Liebman and A. Greenberg, *Strained Organic Molecules,* Academic Press, New York, 1978.

1b. F. Vögtle, *Fascinating Molecules in Organic Chemistry,* J. Wiley, Chichester, 1992.

2. P. E. Eaton, *Angew. Chem., Int. Ed. Engl.* **31** (1992) 1421.

3. K. B. Wiberg, *Chem. Rev.* **89** (1989) 997 and references cited therein.

4. B. R. Venepalli and W. C. Agosta, *Chem. Rev.* **87** (1987) 399.

5. P. E. Eaton and B. D. Leipzig, *J. Am. Chem. Soc.* **105** (1983) 1656.

6. H. Dodziuk, *Top. Stereochem.* **21** (1994) 351.

7. K. B. Wiberg, J. E. Hiatt, and G. J. Burgmaier, *Tetrahedr. Lett.* (1968) 5855.

8a. H. Dodziuk, *J. Mol. Struct.* **239** (1990) 167.

8b. H. Dodziuk and K. Lipkowitz, a communication to the 2nd WATOC Congress, Toronto, Canada, July 1992.

9a. M. P. McGrath, H. F. Schaefer, and L. Radom, *J. Org. Chem.* **57** (1992) 4847.

9b. M. P. McGrath, L. Radom, *J. Am. Chem. Soc.* **115** (1993) 3320.

10. K. B. Wiberg, W. P. Dailey, and F. H. Walker, *J. Am. Chem. Soc.* **105** (1983) 1227.

11. K. B. Wiberg, W. P. Dailey, F. H. Walker, S. T. Waddell, L. S. Crocker, and M. Newton, *J. Am. Chem. Soc.* **107** (1985) 7247.

12. P. Politzer and K. Jayasuriya, *J. Mol. Struct.* **135** (1986) 245; A. B. Pierini, H. F. Reale, and J. Medrano, *J. Mol. Struct.* **148** (1986) 109.

13. P. Chakrabati, P. Seiler, J. D. Dunitz, A.-D. Schlüter, and G. Szeimies, *J. Am. Chem. Soc.* **103** (1981) 7378.

14. P. Dowd and H. Irngartinger, *Chem. Rev.* **89** (1989) 985.

15. The deformation density map is calculated by subtracting from an experimental electron density map the map of electronic densities calculated for a system of isolated atoms located at the corresponding atomic positions of the molecule under study.

16. K. B. Wiberg, S. T. Waddell, and R. E. Rosenberg, *J. Am. Chem. Soc.* **112** (1990) 2184; P. Seiler, *Helv. Chim. Acta* **73** (1990) 1574.

17. D. Feller and E. R. Davidson, *J. Am. Chem. Soc.* **109** (1987) 4133.

18. K. B. Wiberg, R. F. W. Bader, and C. D. H. Lau, *J. Am. Chem. Soc.* **109** (1987) 985, 1001.

19. R. F. W. Bader, P. L. A. Popelier, and T. A. Keith, *Angew. Chem., Int. Ed. Engl.* **33** (1994) 620.

20. H. Dodziuk, *Tetrahedron* **44** (1988) 2951.

21. L. A. Paquette, H. P. Park, and P. F. King, *J. Chem. Res. (S)* (1980) 296.

22. K. Hassenrück, J. G. Radziszewski, V. Balaji, G. S. Murthy, A. J. McKinley, D. E. David, V. M. Lynch, H.-D. Martin, and J. Michl, *J. Am. Chem. Soc.* **112** (1990) 873; D. A. Hrovat and W. T. Borden, *J. Am. Chem. Soc.* **112** (1990) 875; P. E. Eaton and J. Tsanaktsidis, *J. Am. Chem. Soc.* **112** (1990) 876.

23. E. T. Seidl and H. F. Schäfer III, *J. Am. Chem. Soc.* **113** (1991) 1915.

24. R. Hoffmann, R. W. Alder, and C. F. Wilcox, Jr., *J. Am. Chem. Soc.* **92** (1970) 4992.

25. J. M. Schulman, M. L. Sabio, and R. L. Disch, *J. Am. Chem. Soc.* **105** (1983) 743.

26. V. B. Rao, C. F. George, S. Wolff, and W. C. Agosta, *J. Am. Chem. Soc.* **105** (1983) 5732.

27. H. Sachse, *Berichte* **23** (1890) 1363; *Z. Phys. Chem.* **10** (1892) 203.

28. Ch. Kabuto, M. Yagihara, T. Asso, and Y. Kitahara, *Angew. Chem., Int. Ed. Engl.* **12** (1973) 836; W. Littke and U. Drück, *Angew. Chem., Int. Ed. Engl.* **13** (1974) 539; E. Vogel, A. Breuer, C.-D. Sommerfield, E. Davies, and L. K. Liu, *Angew. Chem., Int. Ed. Engl.* **16** (1977) 169; R. Schwesinger, K. Piontek, W. Littke, and H. Prinzbach, *Angew. Chem., Int. Ed. Engl.* **24** (1985) 318 and references cited therein.

29. M. Engelhard and W. Lüttke, *Angew. Chem., Int. Ed. Engl.* **11** (1972) 310.

30. C. Krüger and P. J. Roberts, *Cryst. Struct. Commun.* **3** (1974) 459.

31. E. Osawa and H. Musso, *Angew. Chem., Int. Ed. Engl.* **22** (1983) 1.

32. N. L. Allinger and P. E. Eaton, *Tetrahedron Lett.* **24** (1983) 3697.

33. H. Dodziuk and K. Nowinski, *Bull. Pol. Acad. Sci., Chem.* **35** (1987) 195.

34. R. L. Disch and J. M. Schulman, *J. Am. Chem. Soc.* **112** (1990) 3377.

35. W. Spielmann, H.-H. Fick, L.-U. Meyer, and A. de Meijere, *Tetrahedr. Lett.* (1976) 4057; D. Kaufmann, H.-H. Fick, O. Schallner, W. Spielmann, L.-U. Meyer, P. Golitz, and A. de Meijere, *Chem. Ber.* **116** (1983) 587.

36. J. Spanget-Larsen and R. Gleiter, *Angew. Chem., Int. Ed. Engl.* **17** (1978) 441.

37. J. M. Schulman, M. A. Miller, and R. L. Disch, *J. Mol. Struct. THEOCHEM* **169** (1988) 563.

38. G. M. Kellie and F. G. Riddell, *Top. Stereochem.* **8** (1974) 225; A. N. Vereshchagin, *Usp. Khim.* **52** (1983) Chapters 3, 4.

39. H. Dodziuk, *Bull. Chem. Soc. Japan* **60** (1987) 3775.

40. D. Wehle, N. Schormann, and L. Fitjer, *Chem. Ber.* **121** (1988) 2171.

41. M. Banciu, C. Popa, and A. T. Balaban, *Chem. Scr.* **24** (1984) 28.

42. G. Maier, S. Pfriem, U. Schäfer, and R. Matusch, *Angew. Chem., Int. Ed. Engl.* **17** (1978) 520.

43. T. J. Katz and N. Acton, *J. Am. Chem. Soc.* **95** (1973) 2738.

44. P. E. Eaton and T. W. Cole, Jr., *J. Am. Chem. Soc.* **86** (1964) 962, 3157.

45. L. Cassar, P. E. Eaton, and J. Halpern, *J. Am. Chem. Soc.* **92** (1970) 3515.

46. C. H. Rucker and H. Prinzbach, *Tetrahedr. Lett.* **27** (1986) 1565.

47. P. E. Eaton, Y. S. Or, and S. J. Branca, *J. Am. Chem. Soc.* **103** (1981) 2134.

48. W. Spielmann, H.-H. Fick, L.-U. Meyer, and A. de Meijere, *Tetrahedron Lett.* (1976) 4057; D. Kaufmann, H.-H. Fick, O. Schallner, W. Spielmann, L.-U. Meyer., P. Gölitz, and A. de Meijere, *Chem. Ber.* **116** (1983) 587.

49. C.-H. Lee, S. Liang, T. Haumann, R. Boese, and A. de Meijere, *Angew. Chem., Int. Ed. Engl.* **32** (1993) 559.

50. R. J. Ternansky, D. W. Balogh, and L. Paquette, *J. Am. Chem. Soc.* **104** (1982) 4503.

51. R. J. Brousseau, Ph.D. Thesis, Harvard University, Diss. Abstr. Int. **B37** (1977) 6123b; E. Vedejs and R. A. Shepherd, *J. Org. Chem.* **41** (1976) 742; E. Vedejs, W. R. Wilber, and R. Twieg, *ibid.* **42** (1977) 401; H. Higuchi, E. Kobayashi, Y. Sakata, and S. Misumi, *Tetrahedron* **42** (1986) 1731.

52a. G. Mehta, S. Padma, E. Osawa, D. A. Barbiric, and Y. Mochizuki, *Tetrahedr. Lett.* **28** (1987) 1295.

52b. G. Mehta, S. Padma, E. D. Jemmis, G. Leela, E. Osawa, and D. A. Barbiric, *Tetrahedr. Lett.* **29** (1988) 1613.

53. I. S. Zefirov, A. S. Kozmin, and A. B. Abramenkov, *Usp. Khim* **47** (1978) 289.

54a. G. Maier, S. Pfriem, U. Schäfer, and R. Matusch, *Ang. Chem.* **90** (1978) 520.

54b. G. Maier and D. Born, *Angew. Chem.* **101** (1989) 1085; *Angew. Chem., Int. Ed. Engl.* **28** (1989) 1050.

55. H. Irngartinger, A. Goldmann, R. Jahn, M. Nixdorf, H. Rodewald, G. Maier, K.-D. Malsch, and R. Emrich, *Angew. Chem.* **96** (1984) 967.

56. T. Loerzer, R. Machinek, W. Lüttke, L. H. Franz, K.-D. Malsch, and G. Maier, *Angew. Chem.* **95** (1983) 914.

57. V. I. Minkin and R. M. Minayev, *Zh. Org. Khim.* **17** (1981) 221.

58. R. L. Disch, J. M. Schulman, and M. L. Sabio, *J. Am. Chem. Soc.* **107** (1985) 1904.

59. O. Ermer, P. Bell, J. Schäfer, and G. Szeimies, *Angew. Chem.* **101** (1989) 503; P. v. R. Schleyer and M. Bremer, *Angew. Chem.* **101** (1989) 1264.

60. R. R. Karl, Jr., Y. C. Wang, and S. H. Bauer, *J. Mol. Struct.* **25** (1975) 17.

61. J. M. Schulman and R. L. Disch, *J. Am. Chem. Soc.* **107** (1985) 5059.

62. P. Aped and N. L. Allinger, *J. Am. Chem. Soc.* **114** (1992) 1.

63. R. Gleiter and B. Treptow, *Angew. Chem.* **102** (1990) 1452.

64. G. W. Griffin and A. P. Marchand, *Chem. Rev.* **89** (1989) 997.

65. E. B. Fleischer, *J. Am. Chem. Soc.* **86** (1964) 3889.

66. A. Almenningen, T. Jonvik, H. D. Martin, and T. Urbanek, *J. Mol. Struct.* **128** (1985) 239.

67. E. Hirota, Y. Endo, M. Fujitake, E. W. Della, P. E. Pigou, and J. S. Chickos, *J. Mol. Struct.* **190** (1988) 235.

68. L. Hedberg, K. Hedberg, P. E. Eaton, N. Nodari, and A. G. Robiette, *J. Am. Chem. Soc.* **113** (1991) 1514.

69. I. Hargittai and M. Hargittai, "The importance of small structural differences," in *Molecular Structure and Energetics*, J. F. Liebman and A. Greenberg, Eds., Vol. 2, VCH Publishers, New York, 1987, p. 1.

70. S. A. Shatokhin, L. A. Gribov, and I. S. Perelygin, *Zh. Strukt. Khim.* **27** (1986) 22.

71. J. M. Schulman and R. L. Disch, *J. Am. Chem. Soc.* **106** (1984) 1202.

72. B. D. Kybett, S. Carroll, P. Natalis, D. V. Bonnell, J. L. Margrave, and J. L. Franklin, *J. Am. Chem. Soc.* **88** (1966) 626.

73a. N. L. Allinger, Y. H. Yuh, and J.-H. Lii, *J. Am. Chem. Soc.* **111** (1989) 8551.

73b. H. Dodziuk, *Bull. Pol. Acad. Sci., Chem.* **38** (1990) 11.

74. R. Gleiter and M. Karcher, *Angew. Chem.* **100** (1988) 851.

75. E. Osawa, J. M. Rudzinski, and Y.-M. Xun, *Struct. Chem.* **1** (1990) 333.

76. K. Hassenrück, J. G. Radziszewski, V. Balaji, G. S. Murthy, A. J. McKinley, D. E. David, V. M. Lynch, H.-D. Martin, and J. Michl, *J. Am. Chem. Soc.* **112** (1990) 873.

77. R. Gilardi, M. Maggini, and P. E. Eaton, *J. Am. Chem. Soc.* **110** (1988) 7232.

78. K. Hassenrück, H.-D. Martin, and R. Walsh, *Chem. Rev.* **89** (1989) 1125.

79. E. Honegger, P. E. Eaton, B. K. R. Shankar, and E. Heilbronner, *Helv. Chim. Acta* **65** (1982) 1982.

80. K.-I. Hirao, Y. Ohuchi, and O. Yonemitsu, *J. Chem. Soc., Chem. Commun.* (1982) 99.

81a. C.-H. Lee, S. Liang, T. Haumann, R. Boese, and A. de Meijere, *Angew. Chem., Int. Ed. Engl.* **32** (1993) 559.

81b. A. de Meijere, private communication.

82. G. Mehta and S. Padma, *J. Am. Chem. Soc.* **109** (1987) 2212.

83. V. P. Reddy and E. D. Jemmis, *Tetrahedr. Lett.* **27** (1986) 3771.

84. G. V. Schriver and D. J. Gerson, *J. Am. Chem. Soc.* **112** (1990) 4723.

85. J. M. Schulman, R. L. Disch, and M. L. Sabio, *J. Am. Chem. Soc.* **108** (1986) 3258.

86. E. D. Jemmis, J. M. Rudzinski, and E. Osawa, *Chem. Express* **3** (1988) 109.

87. M. A. Miller and J. M. Schulman, *J. Mol. Struct. THEOCHEM* **163** (1988) 133.

88. G. Dinsburg (D. Ginsburg), *Nouv. J. Chim.* **6** (1982) 175.

89. W.-K. Li, T.-Y. Luh, and S.-W. Chiu, *Croat. Chim. Acta* (1985) 1; E. Osawa, J. M. Rudzinski, A. Barbiric, and E. D. Jemmis, in *Strain and Its Implications in Organic Chemistry,* A. de Meijere and S. Blechert, Eds., Kluver Academic Publishers, Dordrecht, 1988, p. 259.

90. W. D. Fessner, B. A. R. C. Murty, J. Wörth, D. Hunkler, H. Fritz, H. Prinzbach, W. D. Roth, P. v. R. Schleyer, A. B. McEwen, and W. F. Maier, *Angew. Chem., Int. Ed. Engl.* **26** (1987) 51, 52, 55; R. Pinkos, J.-P. Melder, and H. Prinzbach, *Angew. Chem., Int. Ed. Engl.* **29** (1990) 92.

91. J. C. Galucci, C. W. Doecke, and L. A. Paquette, *J. Am. Chem. Soc.* **108** (1986) 1343.

92. J.-P. Melder, R. Pinkos, H. Fritz, J. Wörth, and H. Prinzbach, *J. Am. Chem. Soc.* **114** (1992) 10213.

93. R. Pinkos, J.-P. Melder, K. Weber, D. Hunkler, and H. Prinzbach, *J. Am. Chem. Soc.* **115** (1993) 7173.

94. D. Kuck and A. Schuster, *Angew. Chem., Int. Ed. Engl.* **27** (1988) 1192.

95. R. Boese, T. Miebach, and A. de Meijere, *J. Am. Chem. Soc.* **113** (1991) 1743.

96. L. Fitjer, M. Giersig, D. Wehle, M. Dittmer, G.-W. Koltermann, N. Schormann, and E. Egert, *Tetrahedron* **40** (1984) 4337.

97. D. Wehle and L. Fitjer, *Angew. Chem., Int. Ed. Engl.* **26** (1987) 130; L. Fitjer and U. Quaback, *Angew, Chem., Int. Ed. Engl.* **26** (1987) 1023.

98. M. Traettenberg, P. Bakken, R. Seip, L. Fitjer, and H.-J. Scheuermann, *J. Mol. Struct.* **159** (1987) 325.

99. A. Otterbach and H. Musso, *Angew. Chem., Int. Ed. Engl.* **26** (1987) 554; V. Biethan, U. van Gizycki, and H. Musso, *Tetrahed. Lett.* (1965) 1477; H. M. Hutmacher, H.-G. Fritz, and H. Musso, *Angew. Chem., Int. Ed. Engl.* **14** (1975) 180; V. T. Hoffmann and H. Musso, *Chem. Ber.* **124** (1991) 103.

100. Z. Goren and S. E. Biali, *J. Am. Chem. Soc.* **112** (1990) 893.

# CHAPTER
# 8

# Molecules with Nonstandard Multiple Bonds and Aromatic Rings

In the classical conformational analysis of alkenes and aromatic rings, the interactions among $\pi$ electrons in these molecules determine the planarity of the molecular fragment of which they are a part. To analyze the ethylene deformation model QC calculations were carried out by Wagner et al. [1]. The distortions from planarity can be twisting or pyramidalization. Such a twisting is present in bifluoronylidene **287** [2], for which nearly planar fluorene groups are twisted by ca. 42° about the central double bond. On the other hand, according to X-ray analysis [3], bianthrone **288** adopts an antifolded structure with the 10, 10′ atoms forming the central double bond slightly pyramidalized in opposite directions. Thus the benzene rings form angles of ca. 40° with the mean plane of the double bond. The distortions can be

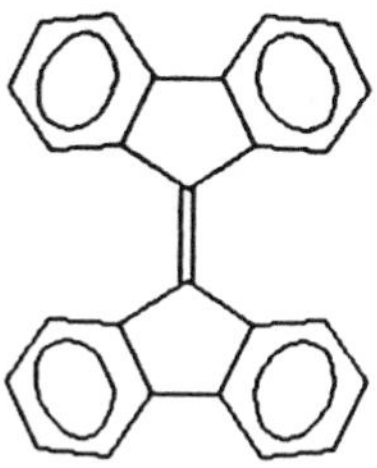

287 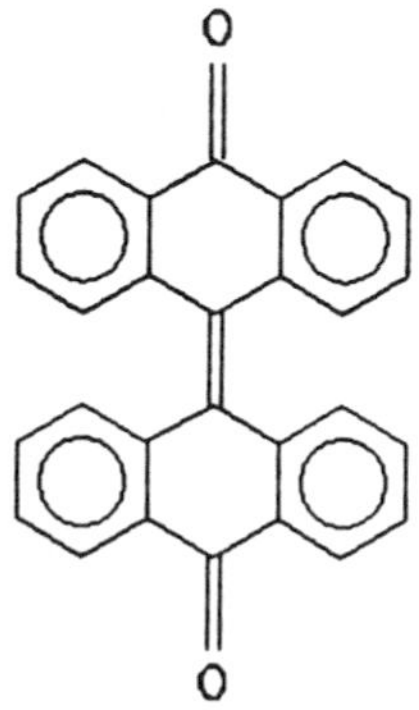                              288

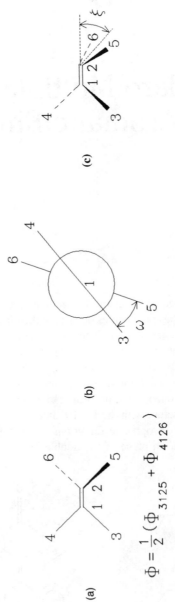

$$\Phi = \frac{1}{2}(\Phi_{3125} + \Phi_{4126})$$

**Figure 8.1** The description of deformations at a double bond. (a) A definition of the mean value of the torsional angle; (b) pure twist; (c) pure pyramidalization.

caused either by steric repulsions of voluminous substituents, or they can be forced by inclusion of a double bond into a medium ring, as in *trans*-cyclooctene **77**. To analyze the degree of the distortion at a carbon atom, the pyramidalization angle (see Fig. 8.1) between the plane defined by the atom C1 and its substituents and the formally double C1C2 bond is introduced. Similarly to the double bond distortions, large ortho substituents at an aromatic ring or the incorporation of the ring into another small- or medium-sized ring causes aromatic ring distortions. An analogous incorporation of a triple bond results in a nonlinear distortion thereof.

## 8.1 Atypical Alkenes [4]

In spite of a strain larger than 50 kcal/mol [5], cyclopropene **289** is planar and isolable, but large substituents on the double bond can enforce its nonplanarity.

**289**

   In spite of attempts to synthesize it, tetra-t-butylethylene **290b** is not known. Tetra-trimethylsilylethylene **290c** has been synthesized [6], but due to a much longer C—Si bond than C—C, it has to be less strained than the former molecule. 1,1,2,2-tetra-N-dimethylaminoethylene **290d** has been obtained. Its X-ray and ED studies have shown [7] that the ethylene bond is twisted by 28°. A similar twisting of 28.6° was found by Krebs et al. for the tetraaldehyde derivative **290e** [8]. The data collected by the latter authors surprisingly reveal that twisting has little effect on the $^{13}$C chemical shifts of olefin protons. The corresponding data for vibrational frequencies put together in Ref. 8 show that the frequency of the Raman stretching vibration is extremely sensitive to the planarity of the double bond. (The Raman spectra, and not IR, are studied here since the band intensity in the latter is prohibitively small.) A typical value for the stretching vibration of planar ethylene **290a** is ca. 1650 cm$^{-1}$, while the lowest value measured for a formally double bond in **290e** shifts to the highly unusual value of 1461 cm$^{-1}$ [8]. The shifts in UV absorption cannot be interpreted in such a simple way.

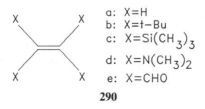

a: X=H
b: X=t–Bu
c: X=Si(CH$_3$)$_3$
d: X=N(CH$_3$)$_2$
e: X=CHO

**290**

As predicted by model calculations [9] and those of higher quality for bicyclo-but-1(3)-ene **291** [10], a *cis* incorporation of a double bond into a small ring results in double bond pyramidalization. Direct experimental evidence for this was found for a highly substituted derivative of bicyclo[$n$.1.0]alk-1($n$+2)-ene **292**, the mole-cule **293** [11]. Maier and Schleyer [12] introduced the concept of "olefinic strain" (OS), equal to the difference in strain energy between an alkene and its correspond-ing alkane hydrogenation product to compare the strain experienced by strained unsaturated hydrocarbons.

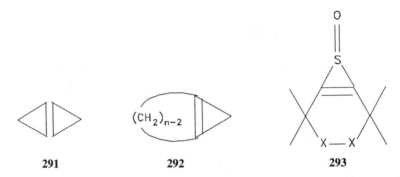

|     291     |     292     |     293     |

A *trans* incorporation of a double bond into a medium ring destabilizes the molecule and forces bond nonplanarity. The calculated olefinic strain is small and negative for *trans*-cyclodecene ($n = 10$) and is zero for *trans*-cyclononene ($n = 9$). The corresponding value for *trans*-cyclooctene ($n = 8$) equals 23.0 kJ/mol (exp. 26.4 kJ/mol), and that for *trans*-cycloheptene ($n = 7$) is 82.1 kJ/mol [12]. **294** with $n = 8$ is isolable and reveals a C—C=C—C torsional angle of 125° [13].

|  $n=8-10$  |  $n=0-3$  |
|  **294**  |  **295**  |

An interesting group of strained cyclopentenes **295** was extensively studied by Borden and Michl et al. [4c, 14, 15] (one member of this group with $n = 1$, **48**, was discussed earlier in Chapter 2). Branan et al. [16] were successful in the synthesis and trapping of the highly strained bis(ethano) derivative of the lowest member of the series with $n = 0$. The compound was calculated to have the very high OS value of 296 kJ/mol and a very weak pyramidalized "double" bond with a dissociation energy of only 55.7 kJ/mol. **295** with $n = 0$ probably has one of the most pyra-midalized formal double bonds. **295** with $n = 1$ (compound **48** from Chapter 2) is

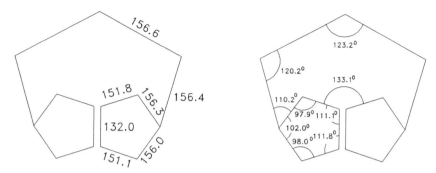

**Figure 8.2**   The geometrical parameters of **295** ($n = 3$) calculated by Hrovat and Borden [15].

also unstable [14]. QC calculations [15] predict a nearly tetrahedral arrangement at each of the "doubly" bonded carbon atoms. As discussed earlier, a weak IR band is present at 1496 cm$^{-1}$ in the matrix isolated spectrum at 10 K of **48**, that is, **295** with $n = 1$. An analogous compound with $n = 2$ was also isolated and characterized in a matrix [17]. A selenium-containing **296** with a longer bridge ($n = 3$) was stable. The latter compound is the only member of the series for which the geometry was established by means of X-ray analysis [18]. Hrovat and Borden carried out ab initio calculations at the 6-31G* level for geometries optimized at the 3-21G level for the **295** series to estimate OS and determine the geometry [15]. The results for $n = 3$ are given in the Fig. 8.2. In view of the lack of corresponding experimental data, their calculated values could not be checked against experiments. This would be especially important for unusually long bonds in the interring bridges and the unexpectedly large value of 123.2° of the central $C_{sp^3}C_{sp^3}C_{sp^3}$ bond in **295** ($n = 3$).

**296**

The incorporation of a double bond into cage compounds also resulted in highly pyramidalized alkenes, some of which can exist only as transient species. Hypothetical prismenes **297**, **298** were studied theoretically [19]. Cubene **272** [20], homo-cub-4(5)-ene **299** [21a], homocub-1(9)-ene **300** [21b], adamantene **301** [22], and quadricycl-1(7)-enes **302** [23] are short-living species or reaction intermediates, while dodecahedrene **303** [24], dodecahedradienes such as **304** [25], and bissecodo-decahedradienes of the type **305** [26] are stable, and X-ray studies have been reported for their derivatives [27]. **305** can also be treated as an unsaturated deriva-

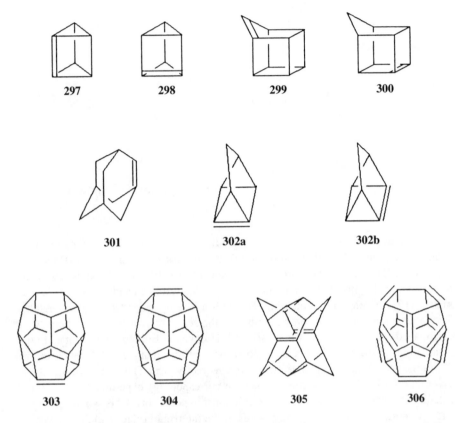

tive of pagodane **278**. The possibility of the existence of fully dehydrogenated $C_{20}$ **306** was discussed by Prinzbach et al. [25, 28].

Smaller torsional distortions were found for the molecules **288** [3] and **307** (13°, [29]), in which the double bond interconnects rings bearing large substituents in the *ortho* positions. The latter molecules belong to the group of *anti*-pyramidalized alkenes.

The double bond is also pyramidalized in *syn*-sesquinorbornene **308** (pyramidalization angle 16–18° [30]) and other simple *syn*-norbornenes.

Highly strained **309** was found to exist in two forms, both involving significantly distorted formal double bonds [31].

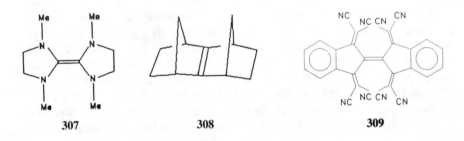

## 8.2 Cyclic Cumulenes

Cumulenes [4d] are the homologous series of hydrocarbons that contain multiple, sequential double bonds sharing a common atom. Allene (1,2-propadiene) **117**, 1,2,3-butatriene **310**, and 1,2,3,4-pentatetraene **311** are the first three members of the series. They have a linear equilibrium arrangement of the skeletal carbon atoms. Cumulenes with an even numer of double bonds with terminal CH bonds lying in perpendicular planes belong to the $\mathbf{D_{2d}}$ symmetry group, whereas those with an odd number of bonds are planar and belong to the $\mathbf{D_{2h}}$ symmetry group. Linear cumulenes are unstrained. They contain standard $sp$- and $sp^2$-hybridized carbon atoms and are highly reactive. Ab initio calculations for bent planar allene yielded an ordering of its possible configurations [32]. Rings larger than ten-membered ones can accommodate the linear allene fragment. Cyclic cumulenes **312** suffer from increasing angle strain as the ring size diminishes from $n = 9$ to 4. The condition of ring closure causes its torsional distortions. The bending of the C1C2C3 angle accompanying the torsional distortions calculated by MNDO QC are given with the formulae **312** [33]. Various computational methods yield similar values for the angles, but strain energies estimated by different methods differ [4d]. Therefore, more accurate estimates of strain energy in cyclic allenes seem desirable.

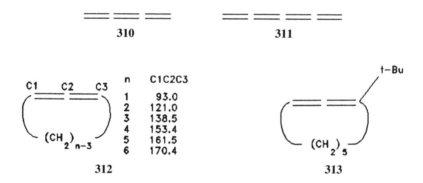

| n | C1C2C3 |
|---|--------|
| 1 | 93.0 |
| 2 | 121.0 |
| 3 | 138.5 |
| 4 | 153.4 |
| 5 | 161.5 |
| 6 | 170.4 |

As mentioned before, 1,2-cyclodecadiene **312** ($n = 10$) is unstrained; thus it is stable, and its allenic fragment is linear [4d]. 1,2-Cyclononadiene ($n = 9$) is easily prepared in multigram quantities but dimerizes on heating. ED [34] and crystal structure [35] determinations for its derivatives yielded a bending of the C=C=C fragment by 10°. The less stable 1,2-cyclooctadiene ($n = 8$) dimerizes faster than the previous one, but it is possible to measure its IR and NMR spectra rapidly [36]. On the basis of semiempirical QC of their normal vibrations [37], the considerable lowering of the IR stretching frequency from 1956 cm$^{-1}$ for 1,2-cyclodecadiene ($n = 10$) to 1550 cm$^{-1}$ for the latter compound ($n = 8$) was interpreted in terms of allene bending in the latter. The 1-tert-butyl derivative **313** seems to be the only isolated compound of this kind [38]. 1,2-Cycloheptadiene **312** ($n = 7$) was found to be too reactive to be isolated. It could not even be observed spectroscopically. An X-ray study of the 1,2-cycloheptadiene–iron complex yielded a structure with a slightly more bent allene fragment than the predicted structure of uncomplexed **312** ($n = 7$) [39].

A chiral allenic structure was assigned to the long-sought **312** ($n = 6$) for a transient state on the basis of STO-3G ab initio calculations [40]. The structure calculated at the 20-configuration MCSCF or MP2 level was also chiral, with a 135° angle of bending and a twisting of the two allenic hydrogens by ca. 30° out of the plane [40]. No definite evidence of 1,2-cyclopentadiene **312** ($n = 5$) seems to have been presented in the literature [4d].

1,2,6,7-Cyclodecatetraene **314** can exist in the form of a *meso* and two D,L structures. An X-ray analysis of the latter yielded an allene fragment bent to 174° [41]. Although the D,L isomers do not appear to be significantly strained [42], they have not been characterized. A chiral bicyclic allene **315** is known [43]. Neither *cis*- nor *trans*-monocyclic 1,2,3-trienes **316a** and **b** were reported, but bicyclic 1,2,3-triene **317** was synthesized, and its NMR and UV spectra were reported [44]. Diketone **318**, containing two 1,2,3-butatriene units, is also known [45].

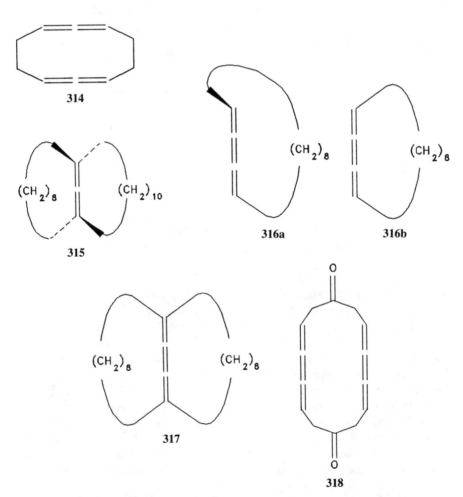

## 8.3 Nonlinear Alkynes [46]

It is much easier to deform a C—C≡C angle than a C—C=C or C—C—C one. As with the cycloalkenes and cyclocumulenes discussed in previous sections, the deformation of a triple bond can be forced by its incorporation into a ring, and the stability of the corresponding compound is determined by the ring size. Cyclononyne **319a** ($n = 9$) and cyclooctyne ($n = 8$) are isolable species; cycloheptyne ($n = 7$), and cyclohexyne ($n = 6$), are short-lived species. Cyclopentyne ($n = 5$), cyclobutyne ($n = 4$), and cyclopropyne ($n = 3$) could not be generated and detected by means of spectroscopic methods even at very low temperatures. However, cyclopentyne derivative **320** was obtained in matrix [47a]. As expected, methyl substitutions in the α positions to the nonlinear triple bond as well as heteroatom substitution with the longer C–heteroatom bond increase the stability of the corresponding cycloalkyne. Introduction of an additional double bond in cyclooctyne increases the strain and destabilizes **319b** [47b,c]. However, two opposing effects are operating in benzannelated cycloalkynes **321**. On one hand, they are destabilized by the increased ring strain. On the other hand, the benzene ring acts as a bulky substituent, kinetically stabilizing **321**. The stabilization effect seems insufficient for cycloalkynes with one benzene ring **321a,** but it causes isolability of **321b** and **321c.** Interestingly, cyclic diyns like **321b** are more stable than the corresponding enynes like **321c.** The effect is understandable, since X-ray analyses revealed that the angles at the triple bond are larger in **321b** [48] than in **321c** [49].

319a

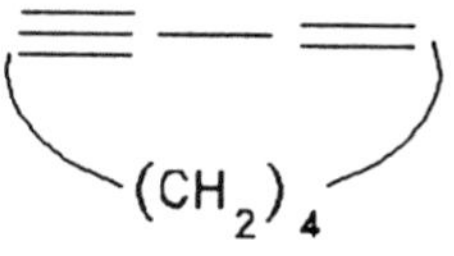

319b

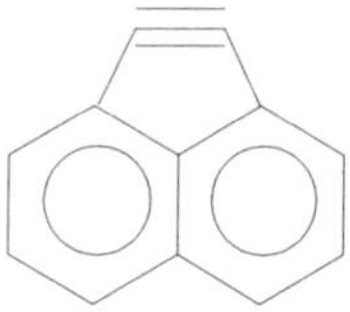

320

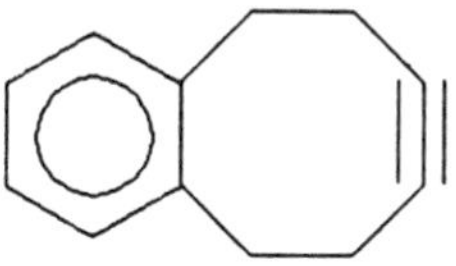

321a

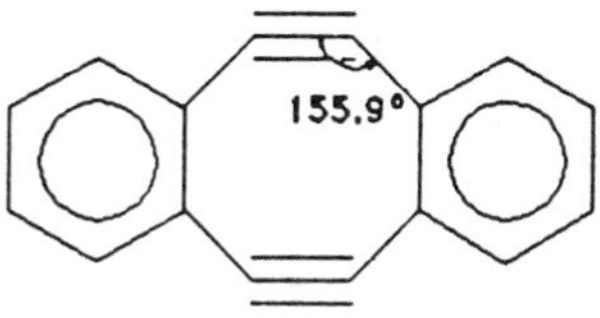

321b 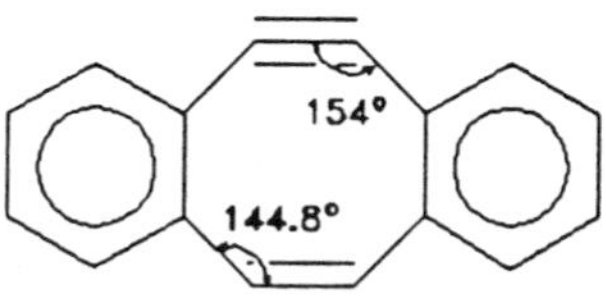     321c

The bending of the $C_{sp^3}C_{sp}C_{sp}$ angle in cycloalkynes from a linear arrangement was determined by means of X-ray or ED techniques to be 160.2° for **319** ($n = 9$) [50], 158.5° for **319** ($n = 8$) [51], 149.3° (the average of 151.7° and 147° for two unequivalent molecules in the unit cell) for **322a** [52], and 145.8° for **322b** [53]. The C≡C bond length changes do not parallel those of the C—C≡C bond angle. The bond length of 123 pm in **319** ($n = 8$) [51] is considerably larger than the value of 121 pm measured in **319** ($n = 9$) [50] and **322b** [53]. In the crystal, 1,5-cyclooctadiyn **323** was found to be almost planar, with the C3—C4 bond elongated to 157 pm and a value of 159.3° for the C≡C—C bond angle [54]. The annelation of two benzene rings in **321b** decreases the angle value to 155.9° [49]. The fusing of two benzene rings to the corresponding 1,5-cyclooctaenyn as in **321c** causes a further decrease in the C≡C—C bond angle to 154°, with an unusually high value of 144.8° for the C=C—C bond angle [48]. The bond angles at the triple bonds in twelve-membered cyclotetraynes **324**, **325** [55, 56], respectively, must be bent by ca. 10–15°.

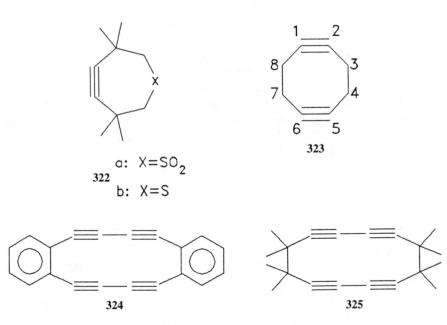

a: X=SO₂

322

b: X=S

323

324                                                                   325

QC and MM calculations [47, 57, 58] reproduce the cited experimental values semiquantitatively, and they provide the values for cycloalkynes that are too unstable to permit measurements or for those for which experimental values are not available. The bond angle values and those of strain energy collected in Table 8.1 reveal the obvious trend: The smaller the angle value, the larger the strain. The experimental values of enthalpies of hydrogenation for some cycloalkynes [47, 57, 58] reflecting the strain are also given in the table. The strain is also reflected in photoelectron [59] and electron transmission spectra [60]. Semiempirical and ab initio QC used to interpret photoelectron spectra of 1,5-cyclooctadiyne **323** and tetracetylene **325** [61, 62] revealed through-space and through-bond interactions between the triple bonds in the molecules. The frequencies and intensity of absorptions in the UV spectra of dibenzocycloakynes **321b,c** were considered as proofs of their planarity [63].

**Table 8.1**  The Calculated Strain Energies, C—C≡C Bond Angle
Values,[a] and Enthalpies of Hydrogenation ($-\Delta H^{25°C}$)
of Angle Strained Cycloalkynes **319**

| Cycle size $n$ | Calc. strain energy (kJ/mol) | Calc. angle (deg) | Enthalpy (kJ/mol) |
|---|---|---|---|
| 9 | 47.3 | 167.4 | 259.4[b] |
| 8 | 80.4 | 158.8 | 296.7[c] |
| 7 | 112.3 | 145.0 | |
| 6 | 191.5 | 130.6 | |
| 5 | 331.8 | 116.2 | |

[a]  D. W. Garrett, Ph.D. Thesis, Indiana University, Bloomington, Indiana, 1974.
[b]  R. B. Turner, A. D. Jarrett, P. Goebel, and B. J. Mallon, J. Am. Chem. Soc. **95** (1973) 790.
[c]  W. R. Roth and H. W. Lennartz, unpublished results cited in Ref. 46.

For obvious reasons the proton chemical shifts are not very sensitive to the triple bond distortions, but the dynamic equilibrium involving the ring inversion in tetramethylcyclooctyne **326** was studied by variable-temperature 60-MHz ¹H NMR. The spectra exhibited only three singlet signals at room temperature. The splittings emerging at 203K allowed one to determine a barrier of 50.3 kJ/mol to the inversion process [64]. The barrier is larger for the dibenzannellated compound **327,** and a coalescence of the $AA'BB'$ signal system of the $CH_2CH_2$ moiety takes place at 128°C. The barrier of 82.5 kJ/mol enabled a chromatographic separation of enantiomers [65].

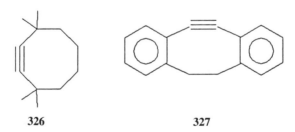

**326**                                            **327**

The effect of ring size on ¹³C spectra is much stronger than that encountered in ¹H spectra. The chemical shift of acetylenic carbon atoms is 94.4 ppm for moderately strained cyclooctyne **319** ($n = 8$) [66], decreases to 87.5 ppm for less strained cyclononyne **319** ($n = 9$) [66], equals 83 ppm for unstrained cyclodecyne **319** ($n = 10$) [47b], and is even lower, 81.5 ppm, for cyclododecyne **319** ($n = 12$) [47b].

C≡C stretching bands are weak in IR and strong in Raman spectra. The smaller the C≡C—C bond angle, the smaller the C≡C stretching frequency [47b]. The effect can be due to the smaller stretching force constant in the deformed bond and/or to the coupling of C≡C stretching to that of the adjacent C—C bond [67].

When the ring size diminishes, the isomeric allenes become more stable than the corresponding cycloalkynes, and the percentage of cycloalkyne to 1,2-cycloalkadiene in the equilibrium mixture increases from 74% for cycloundecyne **319** ($n$ = 11) to 7% for cyclononyne **319** ($n$ = 9) [68].

## 8.4 Cyclophanes

Cyclophanes [69], or shorter phanes, are compounds consisting of one or more aromatic rings connected by one or more bridges. A few examples are given here: [10]metacyclophane **328** ($n$ = 10), [10](1,6)naphthalenophane **329**, ($u,v,w$)meta-cyclophane **330**, [$2_6$](1,2,3,4,5,6)cyclophane **128**, and ($u,v,w$)(1,3,5)cyclophane ($u > v > w$) **331**. The structure, chemistry, and physicochemical properties of cyclophanes, as well as their importance and perspectives, were recently extensively discussed in a monograph by Vögtle [69]. The significance and specificity of this group of compounds were presented in Cram's first paper on cyclophanes [70]. The points of interest raised there are: (1) electronic interactions between "face-to-face" arranged aromatic rings, (2) their influence on substitution reactions in aromatic rings, in particular, the transannular electronic effects induced in one ring by the substituents on another ring, (3) intramolecular charge-transfer complexes, and (4) ring strain, steric strain, and transannular strain. The subsequent rapid development of the field confirmed in detail all these predictions. In addition, the use of macro-

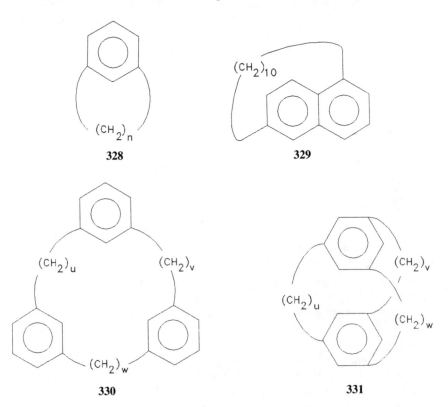

cyclic or polycyclic compounds such as **332** (which in a broad sense also belongs to cyclophanes) in which **333b** fragments play a role of templates self-organizing the formation of topological molecules should be mentioned. The latter fascinating application of cyclophanes will be discussed in the next chapter. In the following we will be interested in the last point raised by Cram, that is, in the structure and strain in cyclophanes as well as in the dynamics of the systems. Bridged aromatic rings in these compounds are often nonplanar and can assume boat, chair, or twist conformations. They are strained, and it influences their spectra and reactivity. Cyclophanes are an extremely large group of compounds, and it is not possible to present them in detail. Therefore, the [$n$]cyclophanes **328** and **334** will be discussed briefly first, then [2.2]paracyclophane **335a,** multiple-layered cyclophanes like **10,** and superphane **128** will be briefly presented. Some other molecules of this type (like **332**), which serve as templates for the synthesis of catenanes and knots, will be mentioned in Section 9.1.

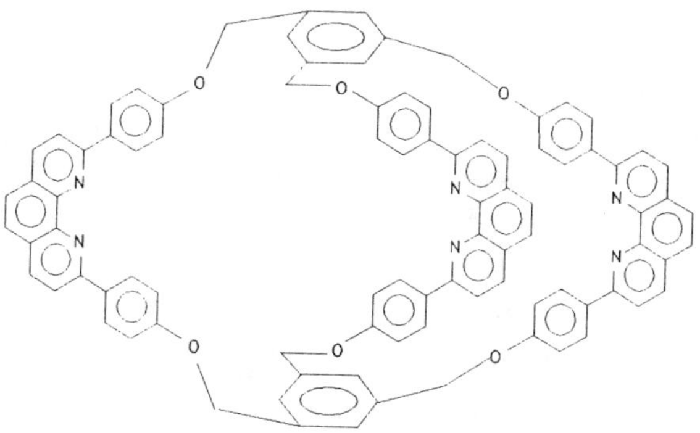

**332**

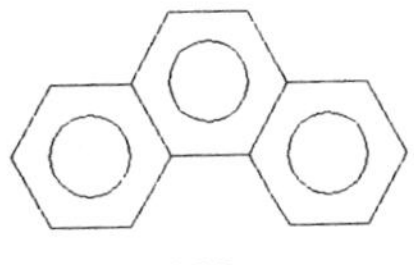

**333a** 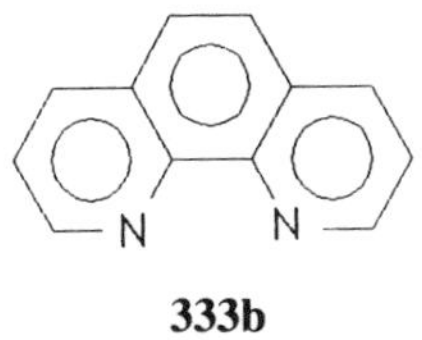          **333b** 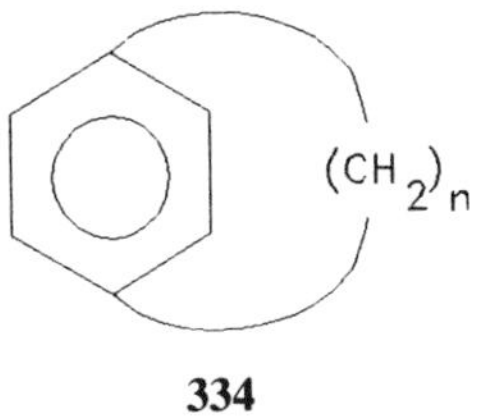          **334**

## 8.4.1 Spectroscopic Properties, the Structure and Calculations for [$n$]Meta- and [$n$]Paracyclophanes

For the [$n$]meta-**328** and [$n$]paracyclophane **334** series it was expected that shorter bridges would lead to a loss of aromaticity as manifested in the UV spectra [71]. This is not the case. With decreasing $n$ values, the bands exhibit a batochromic shift and lose their fine structure. The effect was interpreted in terms of the nonplanar distortions of the aromatic rings without a significant change of aromaticity [72].

$^1$H NMR spectra of [$n$]cyclophanes often exhibit high-field shifts of those aliphatic protons of the bridge that are situated below or above the ring. One example of this kind, **60,** was presented in Section 2.3.8.3. The magnetic anisotropy of the ring can cause considerable shifts. For instance, for 13-bromosubstituted [7]metacyclophane **328** ($n = 7$) there is a signal with $\delta < O$, that is, on the other side of that of TMS [73]. For [6]- and for larger metacyclophanes the bridges undergo a process of pseudorotation. In addition, they can "jump" up and down the ring, executing ring inversion. As a rule the two processes are characterized by different barriers. Thus, for [6]metacyclophane **328** ($n = 6$) the pseudorotation of the methylene bridge has a barrier of 46.6 kJ/mol (the temperature of coalescence $T_c$ equal to $-31.5°C$), while the barrier to ring inversion is much higher ($T_c$ equal to 76.5°C) [73].

X-ray analysis of a carboxyl derivative of [7]paracyclophane **334** ($n = 7$) yielded a boat conformation for the aromatic ring with the $\theta$ angle describing a nonplanar distortion of the ring equal to 17° [74]. A smaller angle of ca. 9° was found for the larger cyclophane with $n = 8$ [75]. MM calculations [76] reproduce these values. For $n \leq 9$, the smaller the bridging ring, the larger the calculated departure from planarity and strain energy [76]. The effect of aromatic ring magnetic anisotropy in [$n$]paracyclophanes **334** was observed in $^1$H NMR spectra [76], while that of pseudorotation of aliphatic bridges in ring-substituted [6]paracyclophanes **334** ($n = 6$) was studied in $^{13}$C NMR spectra [77].

### 8.4.2 [2.2]Cyclophanes

All six possible [2.2]cyclophanes **335** are known [78]. The first one, [2.2]orthocyclophane **335b** is not a typical phane since it does not exhibit through-space interactions of $\pi$-electron systems of its aromatic rings [79]. Evidently, the strongest interaction of this kind manifests itself in [2.2]paracyclophane **335a.** Numerous X-ray studies [80] reveal that the rings are distorted toward the boat conformation with distances between the C3 and C14 atoms of only 275.1–277.8 pm and a C1–C2 bridge distance of 163.0–155.8 pm. A considerable temperature dependence of the bond lengths in **335a** was established, and the effect was interpreted in terms of the strong cooperative movement of the whole system (so-called "concertina movement") discussed in Ref. 80.

The distances between aromatic carbon atoms in **335a** are considerably smaller than the sum of their van der Waals radii. Thus the $\pi$ electrons of the rings form a common system for which the energy difference between HOMO and LUMO orbitals is smaller than the corresponding value of the corresponding open-chain alkylbenzenes [79]. The UV spectrum of [2.2]paracyclophanes is very sensitive to the distances between the ring carbon atoms and to the nonplanar ring distortions [79]. The analysis of the UV and ESR spectra and the photochemical behavior of these molecules by Gleiter et al. [81] showed that bending tightly bound $\pi$ systems such as benzene rings has no significant influence on their ionization energies as long as the deformations do not exceed a critical limit. Substituted [2.2]paracyclophanes were considered to exhibit planar chirality in the Cahn, Ingold, and Prelog classifi-

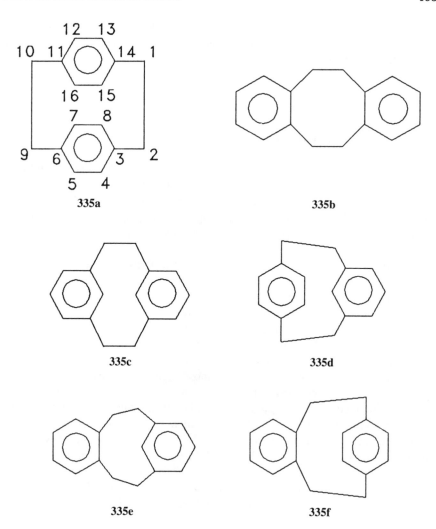

335a

335b

335c

335d

335e

335f

cation [82]. As discussed in Chapter 5, Dodziuk and Mirowicz showed [83] that the axis of chirality and plane of chirality can be replaced by one element, the helical axis. The necessity to replace descriptors of the chirality sense for these systems [84] was also presented there. CD spectra were used to determine the absolute configuration of various [*n*]- and [2.2]cyclophanes [85]. All protons of the aromatic rings are very sensitive to ring substitution of [2.2]paracyclophanes [86] as a result of transannular electronic interactions.

As mentioned before, the interaction between two distorted aromatic rings is especially strong in [2.2]paracyclophane **335a**, layered cyclophanes, the largest known of which is **10,** and superphane **128.** In the latter molecule there is a quite atypical distortion of the aromatic rings, which are planar with all $C_{aryl}C_{alkyl}$ bonds bent toward one side of the rings [87].

Calixarenes **15, 103**, and the molecules **82, 132**, and **332** are also the members of cyclophane family. Most of them are mobile and the symmetry of the average structures of some of them was discussed earlier in Chapter 4.

Diederich [88] recently discussed numerous supramolecular complexes involving cyclophanes as hosts.

## 8.5 Fullerenes [89, 90]

One of the main breakthroughs in chemistry in recent years, buckministerfullerene **11**, holds great promise. Its advent was dubbed the start of the "post-Columbus" chemistry of carbon [90]. Outer- and inner-sphere chemistry with holes or even trap doors in the surface—not to mention a simple trapping of small molecules or ions inside cages—the outlook is exciting. Higher stable fullerenes, $C_{260}$ and $C_{540}$, and the nested, onionlike fullerenes ($C_{60}$ inside a $C_{260}$ cage which, in turn, is inside a $C_{540}$, etc.) were proposed. Several practical applications of these molecules were stipulated. Perfluoration of ball-like fullerenes should yield excellent new lubricants due to their sherical molecular shape. According to many reports, it should be possible to extend the low-temperature superconductivity of fullerenes to higher temperatures, thus diminishing energy losses by the transfer of electricity and revolutionizing our whole life. Holes or trap doors in the fullerene surfaces should enable dialysis at the molecular level and controlled drug release of molecules with pharmacological activity—these are the most exciting practical applications proposed. The first foreseen application of perfluorated fullerenes as lubricants turned out to be impractical, since $C_{60}F_{60}$ was found to be unstable, and, with HF evolved during decomposition [91], it is of no use. Fullerene chemistry is still in the initial stage. For instance, their most promising applications are associated with endo-hedral complexes having an ion inside. Until now, very few fullerene compounds were proved to have such a structure [92]. Only intensive research will show whether the numerous proposals of fullerene applications will be realized or the "fullerene rush" with more than a thousand papers published in 1993 will be just another scientific fad.

The possibility of the existence of molecules with $I_h$ symmetry was questioned until recently (as discussed in Chapter 4), and today there are only two known molecules of such high symmetry. It is interesting that fullerenes are not only obtained artificially but are also found in nature [93]. As shown in Chapter 3, theoretical and synthetic chemistry marched hand in hand, simulating each other in the case of [1.1.1]propellane **5**. On the contrary, $C_{60}$ provides a good example of the separate, thus slower, development of these domains. The molecule was proposed by Osawa and Yoshida [94a] and by Botchvar and Galpern [94b] in the early 1970s on the basis of theoretical considerations, but the idea was totally forgotten. Then, about 15 years later, studies of carbon cluster reactions, aimed at stimulating the conditions under which carbon nucleates in the atmospheres of cool N-type red giant stars, were carried out. Under certain experimental conditions, a cluster with a molecular peak of 720 in its mass spectrum was the most intense. The structure assignment of $C_{60}$ was a tedious task in view of the small amount of the substance available and purification problems. This task would be much easier if it were not carried out independently of the former theoretical ideas [94]. The first proof of the unusual structure of **11** came from its $^{13}C$ NMR spectrum [95]. It contained only one signal, which was compatible only with the high $I_h$ symmetry with all the carbon atoms symmetrically related. Surprisingly, the bond lengths in **11** were determined first by the same techniques [96], since the mobility of spherical full-erene molecules precluded an X-ray determination of its bond lengths in room temperature [97]. The high symmetry of $C_{60}$ imposes restrictions on the vibrational spectra. On the basis of theoretical calculations, it was found that, out of 174 normal vibrational modes giving rise to 42 fundamentals of various symmetries, only four are IR active [98]. The other 10 are Raman active. Having a center of inversion, fullerene **11** provides another example of the operation of the mutual exclusion principle discussed in Sections 2.3.4 and 2.3.5. The nonplanar but at least partly conjugated π-electron system with 60 heavy atoms has been and still is a great challenge for theoreticians. The methods used to study it vary from the HMO [99] to ab initio ones [100] applied to the analysis of heat of formation, electron affinity, ionization potentials, and electronic and vibrational spectra (reviewed in Ref. 89a). Calculations were invoked to explain the dominance of $C_{60}$ and to a lesser extent that of $C_{70}$ in the mixtures obtained by probing laser vaporization of graphite, since these two molecules were found to be the smallest ones in the series that have an isomer in which none of the 12 pentagons abuts [101]. The degree of aromaticity in this group of compounds is of interest, of which the number of possible Kekule structures can be a crude measure. Various theoretical methods reviewed in Ref. 89 have been applied to the problem, and a total of 12,500 structures were calculated. By means of three-dimensional HMO calculations, Fowler and Woolrich [99] pre-dicted that both $C_{60}$ and $C_{70}$ should be closed-shell systems.

An exciting in–out isomerism was postulated for perhydrogenated fullerene $C_{60}H_{60}$. According to the MM calculations [102], the isomer with all CH bonds pointing outward is highly strained in view of the numerous nonbonded interactions and the unfavorable torsional arrangement of neighboring CH bonds. The least strained seems to be the isomer with ten symmetrically distributed CH bonds point-ing into the molecule.

The fullerenes became the name for the whole family of unsaturated hydrocarbon cages $C_n$ built of twelve five- and any number of six-membered rings, but with $n$ assuming only even values. As mentioned, they belong to various symmetry groups, and, for larger $n$ values several isomers of different symmetry are possible [103]. The most stable of them have $n = 60, 260,$ and $540$. The tubes built of the carbon rings, thought to be excellent conductors on one hand, and the strongest fibers known on the other, seem to offer a wide range of new applications [104].

# 8.6 Conclusions

As previously, it should be mentioned that this review of nonstandard unsaturated hydrocarbons is not complete. The field is burgeoning, and the syntheses of numer-ous exciting new molecules are constantly being published. Here, only radialenes

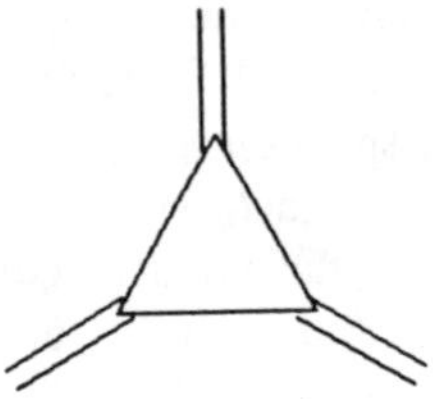

336a

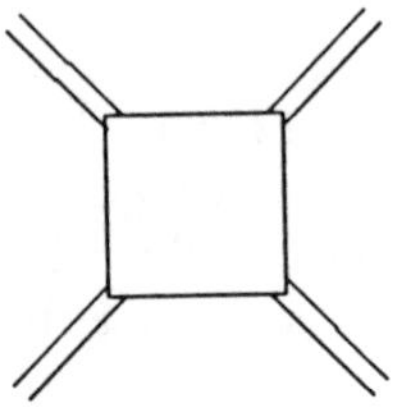

336b

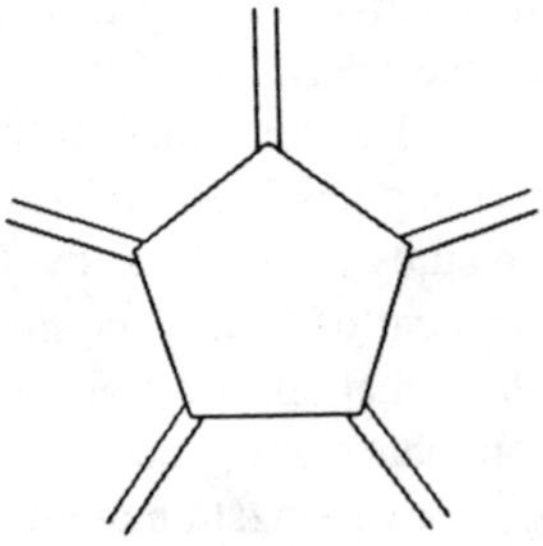

336c

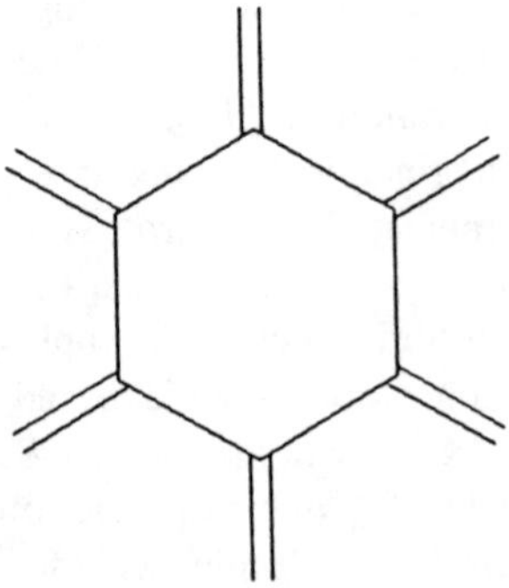

336d

**336, 108,** and **137,** layered cyclophanes like **10** and nonplanar phenanthrenes with voluminous substituents **160** will be mentioned.

Radialenes **336** [105] form an unusual group of compounds having symmetries mostly lower than that suggested by the respective formulae. Some of them were discussed earlier. The sulfur analog of [6]radialene, $C_6S_6$, **49b** with $S_6$ symmetry, discussed in Chapter 2, provides an example of this type with a nonplanar six-membered ring formed by the $C_{sp^2}$ atoms.

Phenantrene **160** ($X = H$) is planar, but its derivatives with bulky substituents $X$ are not [106]. They are chiral, as are helicenes such as **76** discussed in Chapter 4. Diazaphenantrene **333b** is a building block used (e.g., in **332**) by the Dietrich-Buchecker and Sauvage group for the template-directed synthesis of catenanes and knots to be discussed in the next chapter.

## References

1.  H.-U. Wagner, G. Szeimies, J. Chandrasekhar, P. v. R. Schleyer, J. A. Pople, and J. S. Binkley, *J. Am. Chem. Soc.* **100** (1978) 1210.

2.  N. P. Bailey and S. E. Hull, *Acta Cryst.* **B34** (1978) 3289.

3.  E. Harnik and G. M. J. Schmidt, *J. Chem. Soc.* (1954) 3295.

4a.  K. B. Wiberg, *Angew. Chem., Int. Ed. Engl.,* **25** (1986) 312.

4b.  P. M. Warner, *Chem. Rev.* **89** (1989) 1067.

4c.  T. Borden, *Chem. Rev.* **89** (1989) 1095.

4d. R. P. Johnson, *Chem. Rev.* **89** (1989) 1111.

4e. W. Luef and R. Keese, *Top. Stereochem.* **20** (1991) 231.

5. J. F. Liebman and A. Greenberg, *Chem. Rev.* **76** (1976) 311.

6. H. Sakurai, Y. Nakadaira, and H. Tobita, *J. Am. Chem. Soc.* **104** (1982) 300.

7. H. Bock, H. Borrmann, Z. Havlas, H. Oberhammer, K. Ruppert, and A. Simaon, *Angew. Chem., Int. Ed. Engl.* **30** (1991) 1678.

8. A. Krebs, B. Kaletta, W.-U. Nickel, W. Rüger, and L. Tikwe, *Tetrahedron* **42** (1986) 1693 and references cited therein.

9. B. A. Hess, Jr., W. D. Allen, D. Michalska, L. J. Schaad, and H. F. Schaefer III, *J. Am. Chem. Soc.* **109** (1987) 1615.

10. H.-U. Wagner, G. Szeimies, J. Chandrasekhar, P. v. R. Schleyer, J. A. Pople, and J. S. Binkley, *J. Am. Chem. Soc.* **100** (1978) 1210; W. J. Hehre and J. A. Pople, *ibid.* **97** (1975) 6041.

11. W. Ando, Y. Hanyu, T. Takata, and K. Ueno, *J. Am. Chem. Soc.* **106** (1984) 2216.

12. W. F. Maier and P. v. R. Schleyer, *J. Am. Chem. Soc.* **103** (1981) 1891.

13. M. Traetteberg, *Acta Chem. Scand.* **B29** (1975) 29.

14. J. G. Radziszewski, T.-K. Yin, G. E. Renzoni, D. A. Hrovat, W. T. Borden, and J. Michl, *J. Am. Chem. Soc.* **115** (1993) 1454 and references cited therein.

15. D. A. Hrovat and W. T. Borden, *J. Am. Chem. Soc.* **110** (1988) 4710.

16. B. M. Branan, L. A. Paquette, D. A. Hrovat, and W. T. Borden, *J. Am. Chem. Soc.* **114** (1992) 774.

17. T.-K. Yin, F. Miyake, G. E. Renzoni, W. T. Borden, J. G. Radziszewski, and J. Michl, *J. Am. Chem. Soc.* **108** (1986) 3544.

18. D. A. Hrovat, F. Miyake, G. Trammell, K. E. Gilbert, J. Mitchell, J. Clardy, and W. T. Borden, *J. Am. Chem. Soc.* **109** (1987) 5524.

19. V. Jonas and G. Freking, *J. Org. Chem.* **57** (1992) 6085.

20. P. E. Eaton and M. Maggini, *J. Am. Chem. Soc.* **110** (1988) 7230.

21a. D. A. Hrovat and W. T. Borden, *J. Am. Chem. Soc.* **110** (1988) 7229.

21b. N. Chen, M. Jones, Jr., W. R. White, and M. S. Platz, *J. Am. Chem. Soc.* **113** (1991) 4981.

22. R. T. Conlin, R. D. Miller, and J. Michl, *J. Am. Chem. Soc.* **101** (1979) 7637.

23. O. Baumgärtel and G. Szeimies, *Chem. Ber.* **116** (1983) 2180.

24. J. P. Kiplinger, F. R. Tollens, A. G. Marshall, T. Kobayashi, D. R. Lagerwall, L. A. Paquette, and J. E. Bartmess, *J. Am. Chem. Soc.* **111** (1989) 6914.

25. J. P. Melder, R. Pinkos, H. Fritz, and H. Prinzbach, *Angew. Chem., Int. Ed. Engl.* **29** (1990) 95.

26. P. R. Spurr, B. A. R. C. Murty, W.-D. Fessner, H. Fritz, and H. Prinzbach, *Angew. Chem., Int. Ed. Engl.* **26** (1987) 455.

27. M. Keller, K. Scheumann, K. Weber, T. Voss, and H. Prinzbach, *Tetrahedr. Lett.* **35** (1994) 1531.

28. R. Pinkos, J.-P. Melder, K. Weber, D. Hunkler, and H. Prinzbach, *J. Am. Chem. Soc.* **115** (1993) 7173.

29. P. B. Hitchcock, *J. Chem. Soc., Dalton Trans.* (1979) 1314.

30. W. H. Watson, J. Galloy, P. D. Bartlett, and A. A. M. Roof, *J. Am. Chem. Soc.* **103** (1981) 2022.

31. A. Beck, R. Gompper, K. Polborn, and H.-U. Wagner, *Angew. Chem., Int. Ed. Engl.* **32** (1993) 1352.

32. B. Lam and R. P. Johnson, *J. Am. Chem. Soc.* **105** (1983) 7479.

33. R. O. Angus, Jr., M. W. Schmidt, and R. P. Johnson, *J. Am. Chem. Soc.* **107** (1985) 532.

34. M. Traetteberg, P. Bakken, and A. Almenningen, *J. Mol. Struct.* **70** (1981) 287.

35. J. L. Luche, J. C. Damiano, P. Crabbe, C. Cohen-Addad, and C. Lajzerowicz, *Tetrahedron* **33** (1977) 961.

36. J. P. Wisser and J. E. Ramakers, *J. Chem. Soc., Chem. Commun.* (1972) 178.

37. J. P. Price, W. M. Shakespeare, and R. P. Johnson, unpublished results cited in Ref. 4d.

38. J. P. Price and R. P. Johnson, *Tetrahedron Lett.* **27** (1986) 4679.

39. S. M. Oon, A. E. Koziol, W. M. Jones, and G. J. Palenik, *J. Chem. Soc., Chem. Commun.* (1987) 491; F. J. Manganiello, S. M. Oon, M. D. Radcliffe, and W. M. Jones, *Organometallics* **4** (1985) 1069.

40a. M. Balci and W. M. Jones, *J. Am. Chem. Soc.* **102** (1980) 7607.

40b. R. Seeger, R. Krishnan, J. A. Pople, and R. von Schleyer, *J. Am. Chem. Soc.* **99** (1977) 7103.

41. H. Irngartinger and H.-U. Jager, *Tetrahedron Lett.* (1976) 3595.

42. L. Skattebol, *Tetrahedron Lett.* (1961) 167.

43. M. Nakazaki, K. Yamamoto, M. Maeda, O. Sato, and T. Tsutsui, *J. Org. Chem.* **47** (1982) 1435.

44. R. S. Macomber and T. C. Hemling, *J. Am. Chem. Soc.* **108** (1986) 343.

45. P. J. Garratt, K. C. Nicolaou, and F. Sondheimer, *J. Am. Chem. Soc.* **95** (1973) 4582.

46. A. Krebs and J. Wilke, *Top. Curr. Chem.* **109** (1982) 189; W. Sander *Angew. Chem., Int. Ed. Engl.* **33** (1994) 1455.

47a. O. L. Chapman, J. Gano, P. R. West, M. Regitz, G. Maas, *J. Am. Chem. Soc.* **103** (1981) 7033.

47b. H. Petersen, H. Kolshorn, and H. Meier, *Angew. Chem., Int. Ed. Engl.* **17** (1978) 461.

47c. H. Meier, H. Petersen, and H. Kolshorn, *Chem. Ber.* **113** (1980) 2398.

48. R. A. G. de Graff, S. Gortner, C. Romers, H. N. C. Wong, and F. Sondheimer, *J. Chem. Soc., Perkin Trans. II* (1981) 478.

49. R. Destro, T. Pilatti, and M. Simonetta, *J. Am. Chem. Soc.* **97** (1975) 658; R. Destro, T. Pilatti, and M. Simonetta, *Acta Cryst. Sect. B* **44** (1977) 447.

50. V. Typke, J. Haase, and A. Krebs, *J. Mol. Struct.* **56** (1979) 77.

51. J. Haase and A. Krebs, *Z. Naturforsch.* **26a** (1971) 1190.

52. H. H. Bartsch, H. Colberg, and A. Krebs, *Z. Kristallograph.* **156** (1981) 10.

53. J. Haase, and A. Krebs, *Z. Naturforsch.* **27a** (1972) 624.

54. C. Römming, unpublished results cited in Ref. 46.

55. W. K. Grant and J. C. Speakman, *Proc. Chem. Soc.* (1959) 231.

56. R. Weiss, Ph.D. Thesis, University of California, Los Angeles, 1976.

57. N. L. Allinger and A. Y. Meyer, *Tetrahedron* **31** (1975) 1807.

58. D. W. Garrett, Ph.D. Thesis, Indiana University, Bloomington, IN, 1974.

59. H. Schmidt, A. Schweig, and A. Krebs, *Tetrahedr. Lett.* (1974) 1471; P. Carlier, J. E. Dubois, P. Masclet, and G. Mouvier, *J. Electron Spectrosc.* **7** (1975) 55.

60. A. Krebs, W. Rüger, L. Ng, and K. D. Jordan, *Bull. Soc. Chim. Belg.* **91** (1982) 363; L. Ng, K. D. Jordan, A. Krebs, and W. Rüger, *J. Am. Chem. Soc.* **104** (1982) 7414.

61. C. Santiago, K. N. Houk, D. J. De Cicco, and L. T. Scott, *J. Am. Chem. Soc.* **100** (1978) 692.

62. G. Bieri, H. Heilbronner, E. Kloster-Jensen, A. Schmelzer, and J. Wirtz, *Helv. Chim. Acta* **57** (1974) 1265.

63. H. N. C. Wong and F. Sondheimer, *Tetrahedr.* **37** (1981) 99.

64. Ref. 46, p. 215.

65. A. Krebs, *Tetrahedr. Lett.* (1968) 4511; A. Krebs, J. Oldenthal, and H. Kimling, *ibid.* (1975) 4663; H. Meier, H. Gugel, and H. Kolshorn, *Z. Naturforsch.* **31b** (1976) 1270.

66. A. Krebs and H. Kimling, *Liebigs Ann. Chem.* (1974) 2074.

67. L. J. Bellamy, *Advances in Infrared Group Frequencies,* Methuen, London, 1968, p. 21.

68. W. R. Moore and H. R. Ward, *J. Am. Chem. Soc.* **85** (1963) 86.

69a. F. Vögtle, Cyclophan-Chemie, Teubner Studienbücher Chemie, Stuttgart, 1990, pp. 70 ff.

69b. *Ibid.,* p. 93.

69c. *Ibid.,* p. 310.

70. D. J. Cram and H. Steinberg, *J. Am. Chem. Soc.* **73** (1951) 5691.

71. D. J. Cram, C. S. Montgomery, and G. R. Knox, *J. Am. Chem. Soc.* **88** (1966) 515.

72. F. Vögtle and P. Neumann, *Tetrahedron Lett.* (1970) 115.

73. S. Hirano, H. Hara, T. Hiyama, S. Fujita, and H. Nozaki, *Tetrahedron* **31** (1975) 2219.

74. J. Liebe, Ch. Wolff, C. Krieger, J. Weiss, and W. Tochtermann, *Chem. Ber.* **118** (1985) 4144; *ibid.* **118** (1985) 3287; *ibid.,* **122** (1989) 1653.

75. M. G. Newton, T. J. Walter, and N. L. Allinger, *J. Am. Chem. Soc.* **95** (1973) 5652.

76. S. M. Rosenfeld and K. A. Choe, in *Cyclophanes,* Vol. 1, P. M. Keehn and S. M. Rosenfeld, Eds., Academic Press, New York, 1988, pp. 311 ff.

77. J. Leibe, C. Wolff, and W. Tochtermann, *Tetrahedron Lett.* **23** (1982) 2439.

78. Y. Tobe, M. Kawaguchi, K. Kakuichi, and K. Naemura, *J. Am. Chem. Soc.* **115** (1993) 1173.

79. V. Boeckelheide, *Top. Curr. Chem.* **113** (1983) 87.

80. Ref. 69, pp. 92, 93; K. Lonsday, H. J. Milledge, and K. V. K. Rao, *Proc. Roy. Soc.* **A555** (1960) 82; H. Hope, J. Bernstein, and K. N. Trueblood, *Acta Crystallogr.* **B28** (1972) 1733.

81 R. Gleiter, H. Hopf, M. Eckert-Maksic, and K.-L. Noble, *Chem. Ber.* **113** (1980) 3401.

82. R. S. Cahn, C. K. Ingold, and V. Prelog, *Angew. Chem. Int. Ed. Engl.* **5** (1966) 385.

83. H. Dodziuk and M. Mirowicz, *Tetrahedr. Assym.* **1** (1990) 171.

84. K. Schlögel, *Top. Curr. Chem.* **125** (1985) 29.

85. M. J. Nugent and O. E. Weigang, *J. Am. Chem. Soc.* **91** (1969) 4556; W. Tochtermann, U. Vagt, and G. Snatzke, *Chem. Ber.* **118** (1985) 1996.

86. Ref. 69, p. 95.

87.   A. W. Hanson and T. S. Cameron, *J. Chem. Res.* (S) (1980) 336; (M) (1980) 4201.

88.   F. Diederich, *Cyclophanes,* The Royal Society of Chemistry, Cambridge, 1991.

89.   H. W. Kroto, A. W. Allaf, and S. P. Balm, *Chem. Rev.* **91** (1991) 1213; H. W. Kroto, *Angew. Chem., Int. Ed. Engl.* **31** (1992) 111.

90.   J. F. Stoddart, *Angew. Chem., Int. Ed. Engl.* **30** (1991) 70.

91.   R. Taylor, A. G. Avent, T. J. Dennis, J. P. Hare, H. W. Kroto, D. M. R. Walton, J. H. Holloway, E. G. Hope, and G. J. Langley, *Nature* **355** (1992) 27.

92.   M. Saunders, H. A. Jimenez-Vazquez, R. J. Cross, and R. J. Poreda, *Science* **259** (1993) 1428.

93.   P. R. Buseck, S. J. Tsipursky, and R. Hettich, *Science* **257** (1992) 215.

94a.  E. Osawa, *Kagaku (Kyoto)* **25** (1970) 854; Z. Yoshida and E. Osawa, *Aromaticity* (in Japanese), Kagakudojin, Kyoto, 1971.

94b.  D. A. Botchvar and E. G. Galpern, *Dokl. Akad. Nauk SSSR* **209** (1973) 610.

95.   R. Taylor, J. P. Hare, A. K. Abdul-Sada, and H. W. Kroto, *J. Chem. Soc. Chem. Commun.* (1990) 1423.

96.   C. S. Yannoni, P. P. Bernier, D. S. Bethune, G. Meier, and J. R. Salem, *J. Am. Chem. Soc.* **133** (1991) 3205.

97.   W. Krätschmer, L. D. Lamb, K. Fostiropoulos, and D. R. Huffman, *Nature (London)* **347** (1990) 354.

98.   R. E. Stanton and M. D. Newton, *J. Phys. Chem.* **92** (1988) 2141; E. Brensdal, B. N. Cyvin, J. Brunvoll, and S. J. Cyvin, *Spectrosc. Lett.* **21** (1988) 313.

99.   P. W. Fowler and J. Woolrich, *Chem. Phys. Lett.* **127** (1986) 78.

100.  J. M. Schulman, R. L. Disch, M. A. Miller, and R. C. Peck, *Chem. Phys. Lett.* **141** (1987) 45.

101.  T. G. Schmalz, W. A. Seitz, D. J. Klein, and G. E. Hite, *J. Am. Chem. Soc.* **110** (1988) 1113.

102.  Z. Yoshida, I. Dogane, H. Ikehira, and T. Endo, *Chem. Phys. Lett.* **201** (1992) 481.

103.  W. O. J. Boo, *J. Chem. Ed.* **69** (1992) 605.

104.  M. Endo and K. Takeuchi, *Kagaku (Kyoto)* **47** (1992) 710 (in Japanese); X. K. Wang, X. W. Lin, V. P. Dravid, J. B. Ketterson, and R. P. H. Chang, *Appl. Phys. Lett.* **62** (1993) 1881.

105.  H. Hopf and G. Maas, *Angew. Chem., Int. Ed. Engl.* **31** (1992) 931.

106.  A. I. Kitaigorodsky and V. G. Dashevsky, *Tetrahedron* **24** (1968) 5917.

# New Topological Molecules, Starburst Dendrimers, and More

## 9.1 New Topological Molecules: Catenanes, Molecules Mimicking Knots, and Möbius Strips

For a long time mathematicians studied objects of atypical topology [1]. The Möbius strip, schematically presented by formula **30,** is one example of such an object. As illustrated in Fig. 9.1b,c it is formed by a thin strip glued together with one end held rigid and the second one turned around by 180°. Other objects with interesting topological properties are chains of interlocked rings (schematically presented in Fig. 9.2) and knots **13.** (A model of a knot formed by a hydrocarbon chain is presented in color chart 4.) Rotaxanes **31** formed by a ring threaded by a linear fragment with a bulky group at both ends are related to the interlocked rings, but they are isomeric with two separated molecules.

For long time these objects were thought to be toys of mathematicians with no relation to chemistry. Then few molecules representing all these topological types were synthesized. Catenated or knotted structures have also been found in nature [2a]. *Topological isomers* of duplex circular DNA molecules are known, and a whole class of enzymes, *topoisomerases,* effect transformations between them [2b]. [2]- and [3]catenane structures **12, 337,** the knot, and even more complicated topological objects were shown by means of electron microscopy [2] to be mimicked by natural DNA double helices.

The development of synthetic methods for obtaining topological molecules is very interesting [3]. A statistical approach was first applied to the syntheses, which consisted of a standard macrocyclization of a long chain by which a small percentage of rings closed and threaded through another ring formed a [2]catenane **12** [4].

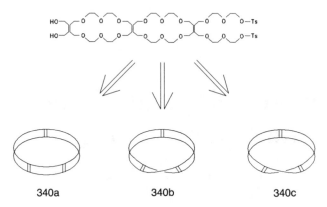

**Figure 9.1**   The scheme of the formation of Walba's ether [2]catenane enantiomers (b and c).

Some catenanes were obtained in this way, but, understandably, the yields of such reactions were low. Much more effective is the templated synthesis of catenands—the metal complexes of the corresponding catenanes proposed by Dietrich-Buchecker, Sauvage, and co-workers [5]. In these reactions the authors made use of a template effect, that is, of the ability of transition metals to gather and dispose ligands around them in a given predictable geometry. The chains that should form the interlocked ring structure contained 1,10-phenanthroline **333b** fragments orienting themselves by complexation with copper ions. Then the free catenane was obtained through demetalation. A series of multicatenands schematically presented in Fig. 9.2 [6], as well as a multiring catenane structure with a bicyclic core **338** (Scheme 9.1) [7], doubly interlocked [2]catenane **339** [8], and a knot **13** [9] were synthesized using this strategy, and an X-ray analysis of the latter was carried out [10]. It should be stressed that without additional elements of chirality catenanes are achiral while Möbius strip [enantiomers depicted in Figs. 9.1(b) and (c)] and knot (Fig. 9.3) are

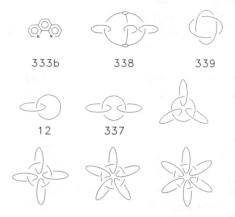

**Figure 9.2**   Polycatenanes synthesized by Dietrich-Buchecker, Sauvage, et al. [6] using **333b** as a template.

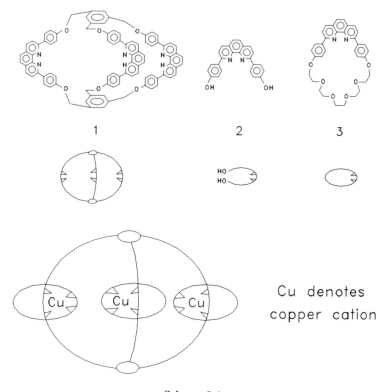

**Scheme 9.1**

chiral. Stoddart and co-workers [11] used another type of preorganization in the synthesis of catenanes producing systems schematically shown in Fig. 9.4, which promise future practical applications.

The first molecule **340b** mimicking a Möbius strip was synthesized by Walba [12]. The topology of a simple cycle with no turn **340a**, that with one turn **340b** = **30,** and the one with two turns is such that breaking ethylene bridges in the first molecule yields two small rings. Breaking the bridges in Walba's ether **30** produces one cycle twice as big as the latter rings, while breaking the C=C bonds in the molecule with two turns gives [2]catenane [4, 13]. The interesting "photoswitchable" catenanes will be presented in Chapter 12 [14]. As mentioned before, catenated and knotted structures have been found in nature [2]. Catenated and knotted DNA have been also synthesized [15].

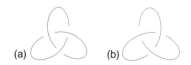

**Figure 9.3** The enantiomers of a knot.

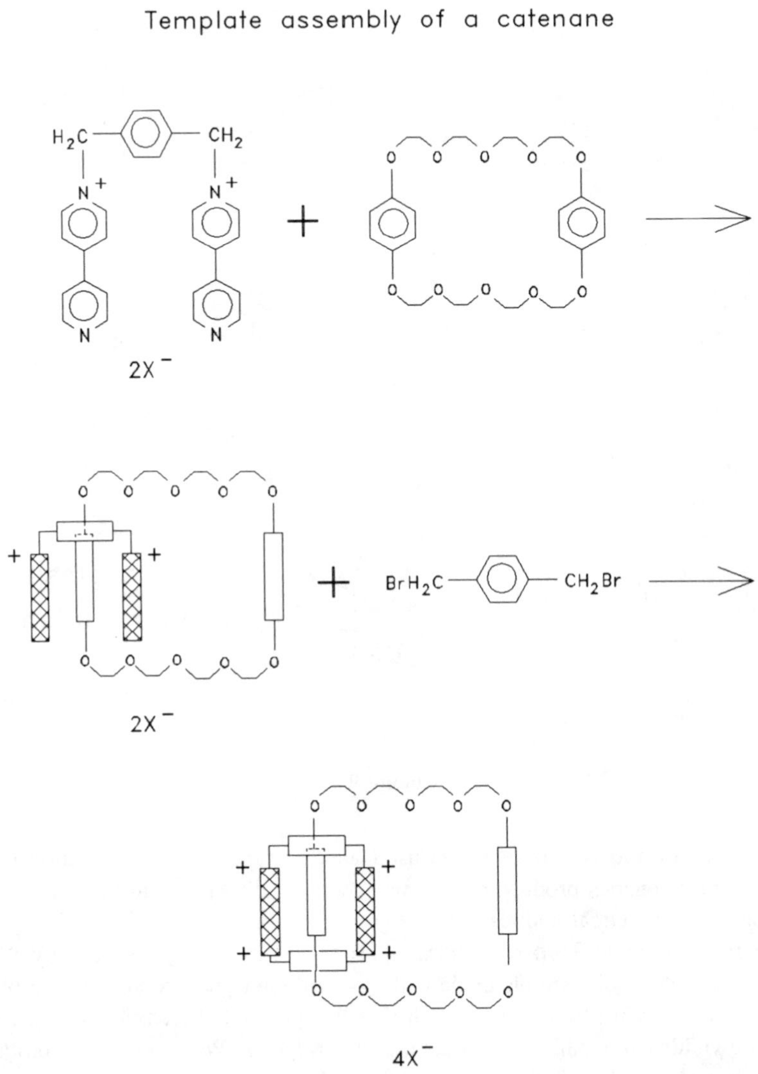

Template assembly of a catenane

**Figure 9.4** The scheme of the [2]catenane synthesis proposed by Stoddart and co-workers [12].

## 9.2 Starburst Dendrimers [16]

Dendrimers [16] or arborols [17] are three-dimensional, highly ordered oligomeric or polymeric compounds formed by reiterative reaction sequences starting from smaller molecules—"initiator cores" such as ammonia [18] or pentaerythritol **341a** [19]. They are "grown" in a controlled way by adding the next generation using

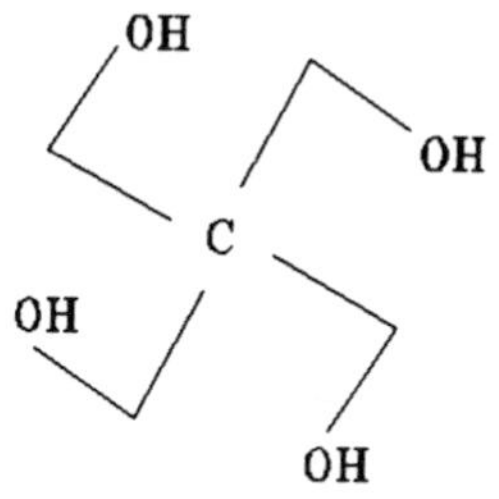

**341a**

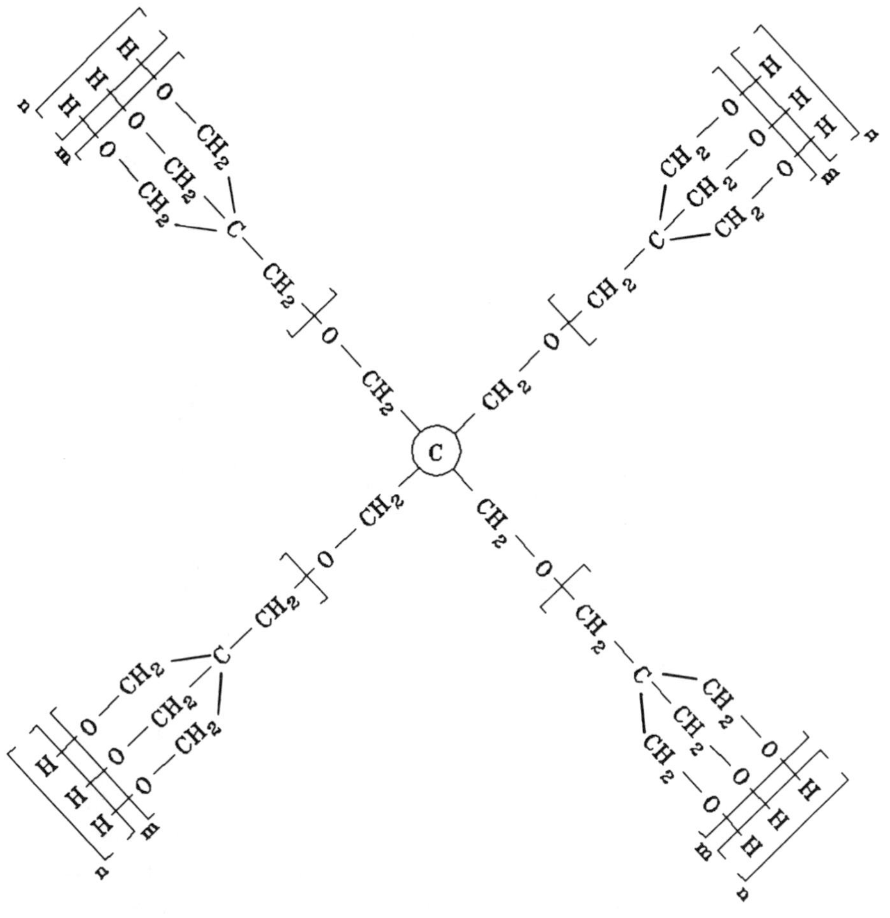

**341b**

subsequent protective/deprotective group strategies. The process involves (1) starting with an initiator core **I** possessing $N_c$ reactive sites, where $N_c$ is 3 in the case of ammonia or the aromatic ring in hydrocarbon dendrimers **342** [20], is 4 for pentaerythritol, etc. (2) A choice of a reaction sequence in which each of the $N_c$

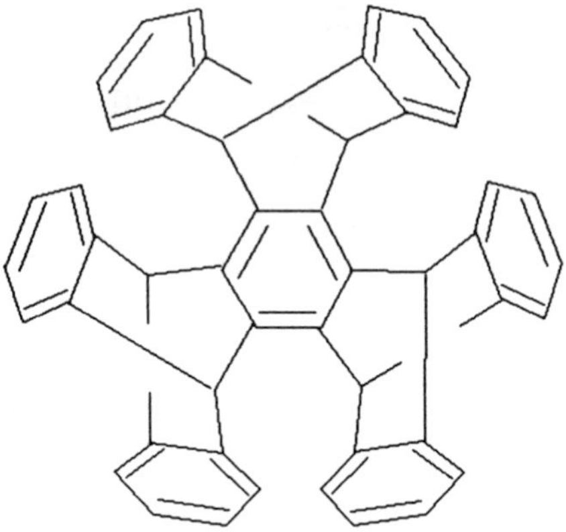

**342a**

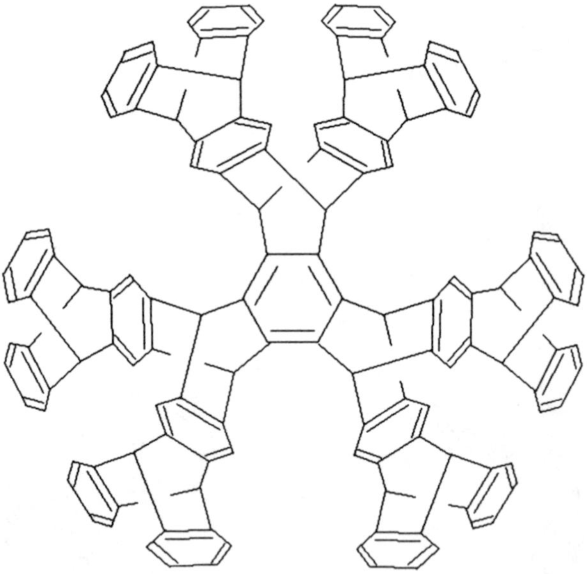

**342b**

reactive sites adds a reactant $B_1$ having $N_b > 1$ new reactive sites. With ammonia as an initiator core, a β-alanine derivative with $N_b = 2$ might be chosen as a reactant, while for hydrocarbon dendrimers **342** and for dendrimers with a penthaerythritol **341a** core, the core initiator and branching fragments are the same. The main problem consists of (3) using protection/deprotection strategies ensuring that $B_1$ reacts with *all* reactive sites of the initiator while no reactions take place at the new reactive sites on $B_1$ of the dendrimer of the zeroth generation $D_o$. Then (4) an iterative sequence of reactions is carried out involving an addition of new reactants

$B_{i+2}$ to the molecule $D_i$ of the $i$th generation to produce a new dendrimer $D_{i+1}$ of the $(i+1)$th generation. In this way the size, shape, topology, flexibility, and surface chemistry of dendrimers can be controlled.

The number of dendrimer terminal groups $z$ is equal to $N_c N_b{}^G$, where $N_c$ and $N_b$ are the multiplicities of the core and the repeating units or branch cells, respectively, and $G$ denotes the generation. The value of $z$ increases rapidly with increasing $G$, and the molecules become almost spherical three-dimensional objects. At higher generations the crowding increases prohibitively, and subsequent ideal growth without defects is impossible.

The unique features of starburst dendrimers situate them between monomeric and polymeric species. Their organization can be studied by graph-theoretical approaches, since treelike dendrimer structures represent Bethe lattices or Cayley trees from a mathematical point of view [21]. For instance, for a given length of the branch segment, core/junction multiplicities, and surface group dimensions of the dendrimer under investigation, one can estimate the starburst limited generation, that is, the biggest possible defect-free generation. Depending on the nature of the core and branching units, the dendrimer interior can be uniformly occupied, or it can contain well-defined channels or cavities, yielding the possibility of an inclusion of smaller molecules. To our knowledge, supramolecular dendrimers with included molecules or ions have not, at present, been explored. However, they seem to offer another challenging possibility of engineering materials exhibiting desired properties.

The structure of starburst dendrimers cannot be studied by X-ray analysis, and their physicochemical studies demand less standard approaches. The most important methods in this field seem to be molecular dynamics simulations and $^{13}$C NMR spectra. The latter technique enables, among other things, a precise detection of defective branches.

As pointed out by Tomalia and co-workers [16], dendrimers can play an important role in mimicking biological receptors. In the future they will also undoubtedly find industrial applications.

## 9.3 Mesoionic Compounds

The domain of nonstandard organic molecules is huge and rapidly developing; their full coverage cannot even be attempted. However, to conclude this section we would like to mention one additional group of compounds.

Mesoionic compounds such as **343** depicted in Fig. 9.5 form a separate group of nonstandard organic molecules. They are built of a five-membered heterocycle that cannot be described by standard covalent or polar structures and have a sextet of $\pi$ electrons and an electronegative substituent [22]. The most interesting results of X-ray analysis [23] of **343**, schematically presented in Fig. 9.5, clearly reveal atypical features pertaining to the carbon atom, which has partly $C_{sp^2}$ character. NMR spectra [24] reflect the charge distribution in these compounds.

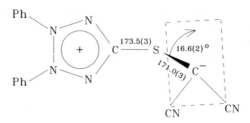

**Figure 9.5** Geometrical parameters of the mesoionic compound **343** [23] (Courtesy of Dr. W. Kozminski).

## References

1. A. F. Möbius, *Gesammelte Werke* (*Collected Works,* in German), Leipzig, 1885; J. C. Chembron, C. O. Dietrich-Buchecker, and J.-P. Sauvage, *Top. Curr. Chem.* **165** (1993) 131, and references cited therein.

2a. E. M. Shekhtman, S. A. Wasserman, and N. R. Cozzarelli, *New J. Chem.* **17** (1993) 757.

2b. J. C. Wang, *J. Mol. Biol.* **55** (1971) 523; M. Gellert, K. Mizuuchi, M. H. O'Dea, and H. A. Nash, *Proc. Natl. Acad. Sci. USA* **73** (1976) 3872.

3. G. Schill, *Catenanes, Rotaxanes and Knots,* Academic Press, New York, 1971.

4. D. M. Walba, *Tetrahedron* **41** (1985) 3161.

5. C. O. Dietrich-Buchecker and J.-P. Sauvage, *Chem. Rev.* **87** (1987) 795.

6. F. Bitsch, C. O. Dietrich-Buchecker, A.-K. Khemiss, J.-P. Sauvage, and A. van Dorsselaer, *J. Am. Chem. Soc.* **113** (1991) 4023.

7. C. O. Dietrich-Buchecker, B. Frommberger, I. Lüer, J.-P. Sauvage, and F. Vögtle, *Angew. Chem., Int. Ed. Engl.* **32** (1993) 1434.

8. J.-F. Nierengarten, C. O. Dietrich-Buchecker, and J.-P. Sauvage, *J. Am. Chem. Soc.* **116** (1994) 375.

9. C. O. Dietrich-Buchecker and J.-P. Sauvage, *Angew. Chem., Int. Ed. Engl.* **28** (1989) 189.

10. C. O. Dietrich-Buchecker, J. Guilhem, C. Pascard, and J.-P. Sauvage, *Angew. Chem., Int. Ed. Engl.* **29** (1990) 1154.

11. P. R. Ashton, T. T. Goodnow, A. E. Kaifer, M. V. Reddington, A. M. Z. Slavin, N. Spencer, J. F. Stoddart, C. Vicent, and D. J. Williams, *Angew. Chem., Int. Ed. Engl.* **28** (1989) 1396.

12. D. M. Walba, R. M. Richards, and R. C. Haltiwanger, *J. Am. Chem. Soc.* **104** (1982) 3219.

13. N. van Gulick, *New J. Chem.* **17** (1993) 619. This article was rejected by *Tetrahedron* in 1960. It was then circulated and cited as a preprint and was found inspiring enough to be published 33 years after its first submission.

14. F. Vögtle, W. M. Müller, U. Müller, M. Bauer, and K. Rissanen, *Angew. Chem., Int. Ed. Engl.* **32** (1993) 1295.

15. S. M. Du and N. C. Seeman, *J. Am. Chem. Soc.* **114** (1992) 9652; J. Chen and N. C. Seeman, *Nature* **350** (1991) 631.

16. D. A. Tomalia, A. M. Naylor, and W. A. Goddart III, *Angew. Chem., Int. Ed. Engl.* **29** (1990) 138.

17. T. Nagasaki, O. Kimura, and S. Shinkai, communication at the XVIII International Symposium on Macrocyclic Chemistry, abstract B-34, Enschede, Netherland, June 27–July 2, 1993.

18. A. M. Naylor and W. A. Goddart, *J. Am. Chem. Soc.* **111** (1989) 2339.

19. H. Hall, A. Padias, R. McConnell, and D. A. Tomalia, *J. Org. Chem.* **52** (1987) 5305.

20. A. Bashir-Hashemi, H. Hart, and D. L. Wart, *J. Am. Chem. Soc.* **108** (1986) 6675; H. Hart, A. Bashir-Hashemi, J. Luo, and M. Meador, *Tetrahedron* **42** (1986) 1641.

21. *Topological Methods in Chemistry*, R. E. Merrifield, and H. E. Simmons, Eds., Wiley, New York, 1989.

22. C. A. Ramsden, in *Comprehensive Organic Chemistry*, P. G. Sammes, Ed., Pergamon Press, 1979, Vol. 4, p. 1171.

23. R. Luboradzki, W. Kozminski, and L. Stefaniak, *J. Crystall. Spectr. Res.* **23** (1993) 133.

24. W. Kozminski, J. Jazwinski, and L. Stefaniak, *J. Mol. Struct.* **295** (1993) 15.

# 10

# Beyond an Isolated Molecule. Complexes with Cyclodextrins and Other Examples of Self-Assembling Objects

## 10.1 Introduction

According to J.-M. Lehn [1a], supramolecular chemistry is chemistry beyond the molecule, the *designed* chemistry of the intermolecular bond. It is a new highly interdisciplinary domain situated among physics, chemistry, biology, and technology. Its aim is to study complexes (supermolecules in [1a]) held together and organized by means of intermolecular noncovalent binding interactions. Lehn [1b] defined supramolecular chemistry as the chemistry of the intermolecular bond, covering structures and functions of the entities formed by the association of two or more chemical species. The delicate balance between rigidity and flexibility was postulated to be crucial for the mutual adaptation of the host and guest and for the stability of the complex formed. The highly selective host–guest interactions provide the basis of self-association processes typical of supramolecular systems, called *self-assembly.*

The specificity of the supramolecular complex formation involves molecular (including chiral, see the following recognition based on the molecular information stored in constituent parts of the complex. The following species are the subject of supramolecular chemistry:

1. Charge-transfer complexes like **344** formed by [2.2]cyclophane with tetra-cyanoethylene [2];
2. Coordination complexes formed by one or more atoms or ions with *n* organic ligands like the complexes of crown ethers **15** with alkaline ions and those mentioned in Section 9.1 on the template synthesis of catenanes and knots [3].

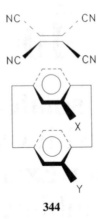

**344**

One example of such a complex reminiscent of a Russian nested doll toy (shown in part in color chart 5) will be presented in the following.

3. The complexes held together by hydrogen bonding (like the one proposed by Rebek that should mimic the self-reproduction of living systems [4] discussed in the Introduction, and beautiful self-assembling structures like **17** and those presented in Scheme 10.1 synthesized by Whitesides et al. [5]).

4. Another type of supramolecular object is the clathrates or the solid-state inclusion compounds, in which the guest molecules enter voids between the hosts in the crystal. This type of supramolecular assembly is presented in Ref. 1c. These four groups of supramolecular complexes will not be discussed here.

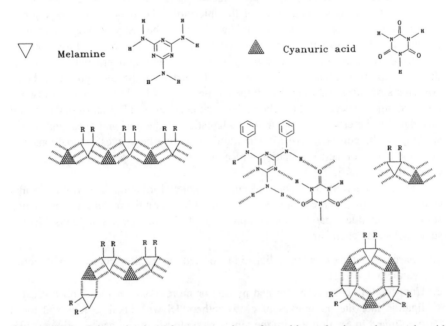

**Scheme 10.1** Schematic view of some complexes formed by melamine and cyanuric acid involving numerous hydrogen bond formations. (Adapted from Ref. 5.)

5. The last, but not least, group consists of the inclusion complexes formed, among others, by hemicarcerands **345** [6] or cyclodextrins **14a–c** [7, 8] with various smaller molecules. In the latter case the larger molecules possessing cavities are called *hosts,* and the smaller ones entering the cavities of the latter to form the

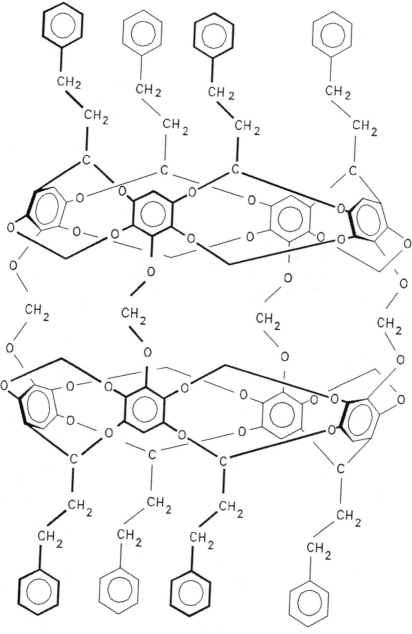

**345**

complex are called *guests*. The complex of **345** with cyclobutadiene **181** is probably one of the best examples of the unique properties of a supramolecular system. Here the hemicarcerand molecule acts as a reaction flask for highly unstable cyclobutadiene, enabling its synthesis and storage at room temperature for a long time.

Why are such objects worth studying? What distinguishes them from typical organic molecules? Supramolecular complexes are interesting from both a theoretical and a practical point of view. First, the complexation process is still not well understood. Several driving forces for the complex formation take part in the process: (1) weak but very numerous van der Waals interactions, (2) electrostatic interactions (interactions between ions, dipole–dipole interactions, charge-transfer interactions, etc.), and (3) hydrogen bonds. The entropy change during the complexation can also play an important role in the process. All these interactions are the most effective when there is a complementarity of the shapes and electronic features of the host and guest molecules and there is an appropriate rigidity–flexibility balance guaranteeing their mutual fitting and the function performed by the complex, which will be discussed later. The "lock-and-key" model of the fit between the molecules was proposed by Fisher as early as in 1894 [9]. The later "induced-fit" model, implying mutual adaptation of both parts of the complex, seems to provide better conditions for the complexation [10].

As discussed in Chapter 1, self-assembly is typical of processes in living nature. It characterizes itself by the very high efficiency and selectivity illustrated there by the decomposition and self-assembly of tobacco mosaic virus briefly presented there. Mimicking the unrivalled effectiveness of the self-assembly processes in nature is one of the aims of studies in this domain. Another is the study of the functioning of self-assembled structures. Yet another is the modeling of enzyme operation to explain the ways in which the complexation process effectively influences the reactivity of the guest molecules. The catalytic amplification of some reactions involving ferrocene by a factor of 750,000 or more [11] by **14b** shows why CyDs serve as enzyme models. Thus the studies of supramolecular complexes should, among other things, provide a better understanding of enzyme catalytic processes and of transport through membranes (playing an extremely important role in living organisms), as well as of molecular recognition and information storage in living organisms. Moreover, most current models of the origin and evolution of life are based on the idea of self-assembling molecular systems [12]. Thus studying such processes allows one not only a better insight into the way living organisms operate, but it also helps to answer more fundamental questions.

Another field of interest in supramolecular chemistry is the industrial use of molecular and chiral recognition processes. It involves chromatographic separations, new modes of drug administration, and the creation of novel recognition-dependent devices in which the specific mode of operation is different from that of the isolated components of the complex. The components may be photoactive, electronoactive, or ionoactive, enabling light conversion by energy transfer [1, 2], photosensitive sensors, creation of new materials exhibiting nonlinear optical effects, organic conductors and semiconductors, and switches. At present the application of supramolecular systems is a vast area of research that will only briefly be discussed in Chapter 12.

## 10.2 Cyclodextrins, Their Properties, and Applications

$\alpha$-, $\beta$-, and $\gamma$-cyclodextrins (CyDs) **14a–c** are cyclic oligosaccharides consisting of six, seven, or eight glucopyranoside units, respectively, interconnected by $\alpha$-(1,4) bonds [7, 8]. They are products of the enzymatic degradation of dextrin. Their average structure is that of a truncated-cone with the approximate dimensions in ppm given in the Fig. 10.1. Sundararajan and Rao have shown theoretically [13] that the numer of glucopyranoside rings $n$ in CyDs cannot be lower than six. Larger CyDs with $n = 9$–12 are known, but, owing to the difficulties in their preparation, they have been studied less and have not found industrial applications.

The structure of CyDs shown in formulae **14** and in color chart 6 implies the existence of a cavity capable of including a guest molecule. According to an adequate formulation by Stoddart, they are "all-purpose molecular containers for organic, inorganic, organometallic, and metalloorganic compounds that may be neutral, cationic, anionic, or even radical" [14]. CyDs are very interesting, both from a theoretical and a practical point of view. They enable the study of the complex formation mechanism and can also serve as models of enzyme catalysts. Their low cost and the ease of highly selective complexation form a basis for their countless practical applications in the pharmaceutical, agrochemical, and tobacco industries, in food, cosmetics, and explosive production, in electronics, and in many other fields. To name but a few, the most important ones that are discussed in detail in Ref. 15 are:

1. Microencapsulation of drugs and herbicides sensitive to light and humidity. It enhances their stability and solubility, thus improving their bioavailability on one hand. On the other, it can reduce irritation caused by the drug, or decrease its bad taste or smell. The preparation of drug mixtures that can react with each other can also be facilitated by complexing the most active agent with CyDs. Their complexes can be used to revolutionize drug administration, since a slow, controlled release of the drug from the complex can replace its intake every few hours.
2. Complexation with CyDs, enhancing stability, is also of importance for explosive substances, making their handling risk-free. Nitroglycerin **346,** both the drug and the explosive, is prepared in this way [16].

$$CH_2ONO_2$$
$$|$$
$$CHONO_2$$
$$|$$
$$CH_2ONO_2$$

**346**

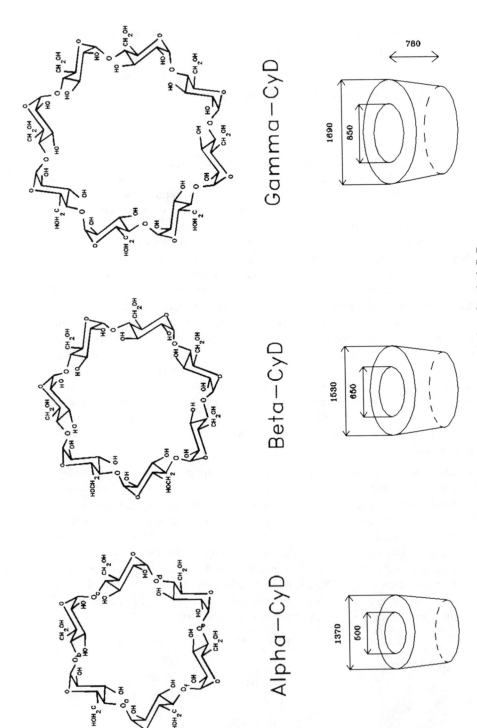

**Figure 10.1** A schematic view and size of typical CyDs.

3. Chromatographic separation of components from complicated mixtures involving:

   a. Removal of undesired components. This may involve the elimination of components responsible for the bitter taste in juices, of those with an unpleasant flavor from cosmetics products, or withdrawal of nicotine and tar from tobacco.

   b. The separation of diastereoisomers and enantiomers. For the latter task, chromatography, making use of CyDs as the mobile or stationary chiral phase [17], is the most effective means. As discussed in Chapter 5, enantiomer separation is especially important for chiral drugs for which one enantiomer can have the desired pharmacological activity, while the second, present as a 50% impurity in the drug, is a ballast at best. At worst, it can exhibit considerable toxicity.

4. The stabilization of the excited states of dyes by CyDs complexation can be used in photoelectronics. This usage will be discussed in general terms in Chapter 12, dealing with applications.

## 10.3  Structure of Cyclodextrins and Their Complexes

CyDs studies are a difficult task since (1) their ease of complexation makes difficult the removal of impurities, and (2) they can form complexes of varying stoichiometry, ternary complexes involving solvents, etc. (3) The complex formation process critically depends on the solvent, concentration, pH, etc. (4) The size of cyclodextrins and the specificity of their energy hypersurface involving numerous low-lying energy minima makes obtaining reliable theoretical results difficult.

Until recently, on the basis of numerous X-ray analyses, CyDs have been considered to have a rigid truncated-cone structure with the primary hydroxyl groups situated at the thinner base of the cone and the secondary ones situated at the broader base [18]. As shown in Fig. 10.2, H3 and H5 protons, as well as the lone

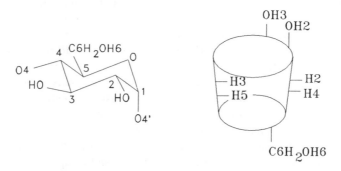

**Figure 10.2**  The atomic numbering and relative positions of CyDs protons.

pairs of interglycosidic oxygen atoms, point inside the cavity, while H2 and H4 protons are located outside the cavity. Such a structure implies a relatively nonpolar character of the cavity. The Lehn notion [1a] on the importance of rigidity versus flexibility balance in the complexation process has not been recognized. Most calculations [19] have been carried out starting from the rigid symmetrical structures with a $C_6$, $C_7$, or $C_8$ symmetry axis for $\alpha$-, $\beta$-, and $\gamma$-CyDs, respectively, or from the rigid experimental structures. The rigidity of the CyDs contradicts the NMR results discussed in Chapter 2 and those to be presented. As will be discussed, CyDs complexes possess a considerable degree of flexibility, even in the solid state, which is larger in solution, where most applications take place [20]. As shown by NMR and X-ray results for para-nitrophenol **37c** ($X = NO_2$) complex with hexakis-(2,3,6-tri-O-methyl)-$\alpha$-CyD **36a** presented schematically in Chapter 2 in Fig. 2.18 [21], the mode of inclusion can be influenced by crystal forces.

The CyDs rigidity is also incompatible with the ease of their complex formation with molecules of different shapes and with model considerations. They are known to be formed by relatively rigid glucopyranoside rings in the $^4C_1$ conformation connected by 1,4 ether linkages. The condition of the macrocycle ring closure and the low barriers to internal rotation around the C—O bond (ca. 4 kJ/mol [22]) make the rigidity of CyDs highly improbable. An inspection of the steric energy surfaces of the $\alpha$-CyD molecule distorted in various ways, presented in Figs. 10.3–10.5, lends credibility to the latter statement. The energy dependence on the inter-glycosidic O···O distance between the opposite oxygen atoms is shown in Fig. 10.3. The distance was scanned from 40 to 1300 pm by 5 pm first, and the mini-mum energy was calculated [23] by MM [24] with MM2 parametrization [25] with the distance constrained. Then the same calculations were repeated for distances changed in the opposite direction from 1300 to 40 pm. As can be seen from a comparison of Fig. 10.3(a) and (b), the two curves are completely different. This is understandable in view of the multidimensionality of the CyD energy hypersurface. The energy dependence on the rotation of a glucopyranoside ring around the $O_a$–$O_b$ axis exhibits several low-lying minima, as does that of the two-dimensional driver, involving simultaneous scanning of the $O_aO_bO_cO_d$ and $O_bO_cO_dO_e$ dihedral angles shown in Figs. 10.4 and 10.5, respectively. The latter image, showing a vast area of acceptable torsional angles within the range of thermal energy at room tempera-tures, clearly exhibits the CyD macrocycle nonrigidity. These model calculations were carried out for an isolated molecule under several simplifying assumptions (e.g., neglect of the solvent or force-field dependence). Nevertheless, they quali-tatively confirm the flexibility of CyDs inasmuch as they do not include entropy effects associated with an $n$-fold degeneracy of the system. Due to the latter, any deformation of one glycopyranoside ring in **14a** has the same probability as the same deformation of any other of these rings, decreasing the entropy by a statistical factor proportional to ln 6 for $\alpha$-CyD, that of ln 7 for $\beta$-CyD, etc.

The dynamic character of the CyDs complexes is clearly revealed by their NMR spectra in solutions. The prevailing majority of them [26] exhibit only one average signal for each proton (or carbon) atom of the free and complexed host and guest

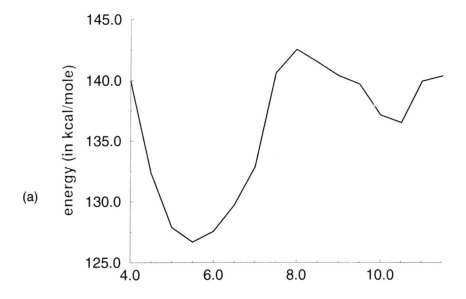

(a)

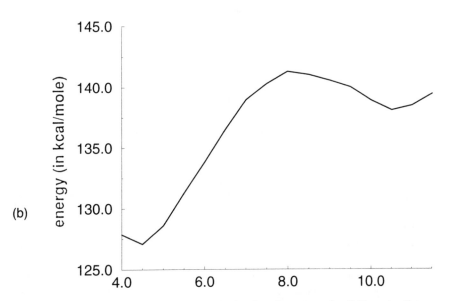

(b)

**Figure 10.3** The dependence of the MM-calculated steric energy of α-CyD on the distance between opposite interglycosidic oxygen atoms. (a) Stretching from 40 to 1300 pm (4 to 13 Å). (b) Compressing from 1300 to 40 pm.

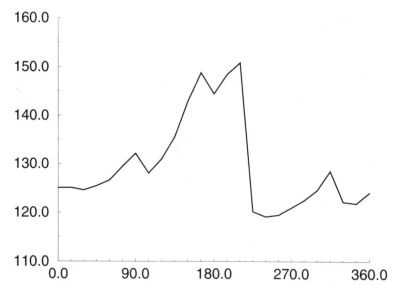

**Figure 10.4**  The dependence of the MM-calculated steric energy of α-CyD on the angle of rotation of one glycosidic unit out of the plane of the interglycosidic oxygen atoms.

**347**

species. Only in the case of rotaxane-type complexes with **14** [27] and of the 1:2 complex of acenaphthene **347** (Fig. 10.6) with β-CyD **14b** at low temperatures [28] have the separate signals been observed due to an increased barrier for the complex formation. The $^1$H spectra of most complexes with CyDs exhibit only one signal of the H1 (and others) protons in the spectrum, evidencing either a completely rigid symmetrical structure or the existence of rapidly exchanging distorted structures. The former hypothesis is improbable in view of the mechanical properties of the systems discussed previously. The lack of splitting of the H3 and H5 signals pointing inside the CyD cavity (see Fig. 10.2 for the atom numbering) in complexes with aromatic hydrocarbons, which should experience the influence of the ring's magnetic anisotropy in the rigid complex, also enhances confidence in the CyDs complex's flexibility. Moreover, the considerable mobility of the complexes is retained even in the solid state. It is shown by the rapid cooperative change of direction of all hydrogen bonds formed by the secondary hydroxyl groups at room temperatures

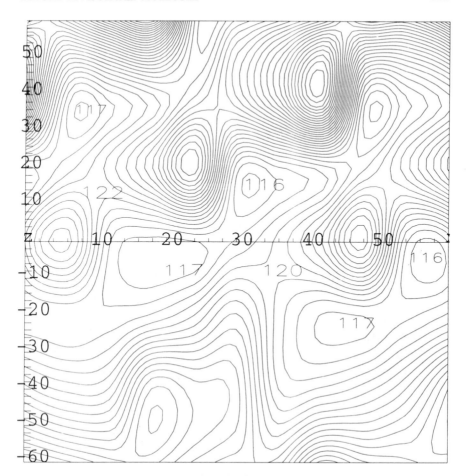

**Figure 10.5** The dependence of the MM-calculated steric energy of α-CyD on the simultaneous dihedral driver of $O_aO_bO_cO_d$ and $O_bO_cO_dO_e$ torsional angles.

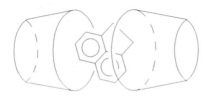

**Figure 10.6** The schematic view of the 1:2 complex of acenaphthene **347** with β-CyD.

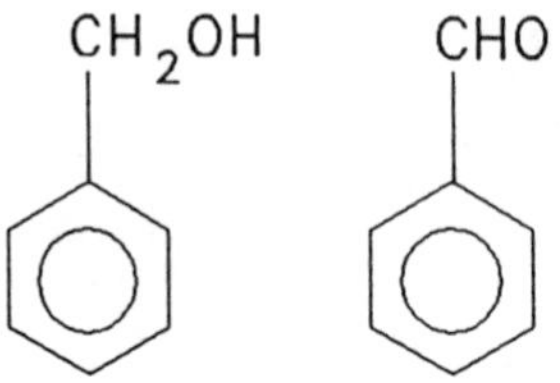

**348**                          **349**

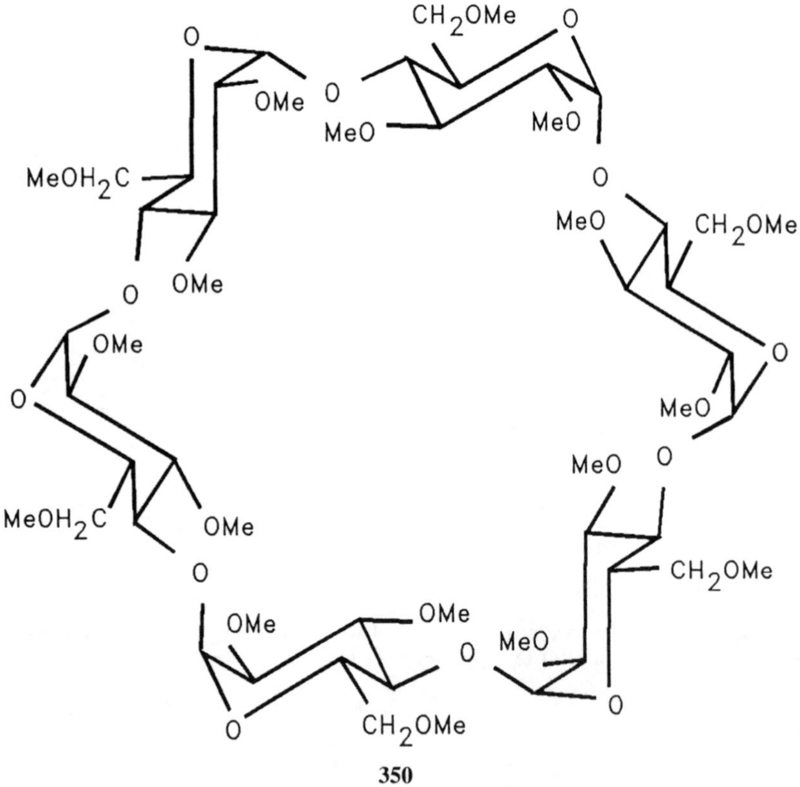

**350**

detected with neutron diffraction of β-CyD discussed in Chapter 2 [29] and in the aromatic ring flop in benzyl alcohol **348** complexes with β-CyD as jointly studied by $^2$H solid-state NMR and X-ray techniques [30]. Similarly, the presence of one signal of the *ortho*-carbon atoms in the $^{13}$C spectra of the complex of benzaldehyde **349** with α-, β-, and heptakis-(2,3,6-tri-O-methyl)-α-CyD **14a**, **14b**, and **350**, respectively, indicates the ring flap [31]. Similarly to the observations in the complex of **348** with β-CyD in the solid state, the flap was undetectable in X-ray analysis, which yielded a static structure of the complex [32].

As stated before, the complexation by CyDs is highly selective. **18** and **347** were

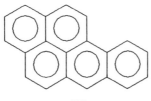

$$R = CH_2C(OH)HCH_2NH_2C_3H_7$$

**Figure 10.7**  A schematic view of a drug complex revealed by NOE.

established chromatographically to form much more stable complexes with β-CyD than all other isomeric dimethylnaphthalenes studied [33]. Its stability constant in a 50% by vol. water/ethanol solution at room temperature was found to be at least 50 times larger than those for other isomers, and the complex was present even at 90°C when all others dissociated into their constituent parts. Excellent chromatographic separations of enantiomers **19, 20** of pinene mentioned in Chapter 1 were also observed [34]. Such highly selective molecular and chiral recognitions by CyDs form the basis of most of their applications. The $^1H$, $^{13}C$, and $^{15}N$ NMR manifestations of the latter are not numerous [28], but, as pointed out in Ref. 36, they can be used to determine the enantiomeric purity of chiral drugs. An example of the $^{13}C$ NMR manifestation of the chiral recognition of enantiomeric invertomers of *cis*-decalin **25** by β-CyD at low temperatures [28] was presented in Chapter 2. The use of NOE to determine the spatial structure of a complex of diphenyl-porphyrine **55** with heptakis-(2,6-di-O-methyl)-β-CyD **36b** was also presented there [37]. The close distance between H3′ and H5′ of CyD and H8 of the drug guest (and that of H5 and H5′), also was determined by NOE, enabled Casy and Mercer to propose a structure for the drug complex depicted in Fig. 10.7. The Job method of determining the complex stoichiometry [38] and that of Benesi and Hildebrand adapted for NMR for a determination of the stability constant of the complex [39] will only be briefly mentioned here.

The inclusion of an achiral guest into the chiral CyD cavity renders it chiral. Out of several examples reported in the literature, an induced CD spectra of otherwise achiral **351** can be mentioned [40].

**251**

The preceding discussion of the properties of CyDs complexes was limited almost exclusively to the CyDs complexes with hydrocarbons since (1) in view of the rapid development in the field, its full coverage is not possible, while (2) the CyDs complexes with hydrocarbons are especially interesting for the study of driving forces for complex formation, since there is no possibility for hydrogen-bond formation, $\pi$ stacking, and/or considerable electrostatic interaction in these systems. Chromatography and NMR spectra seem to be the main tools for studying CyDs complexes in solution and in the solid state. At present only model calculations of limited reliability can be carried out for the complexes. The further development of computers will enable their theoretical study.

The reported data on CyDs complexes are very numerous, and it is not possible to cover this field of research completely. Therefore, to conclude this section, only a few, but interesting, examples of the complexes will be mentioned. A complex of $\gamma$-CyD with [12]crown-4 ether **352** inside the CyD cavity, which in turn is capable

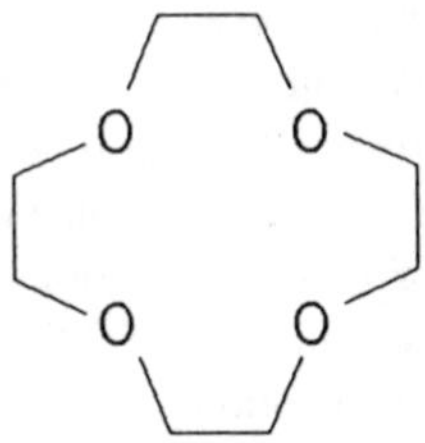

**352**

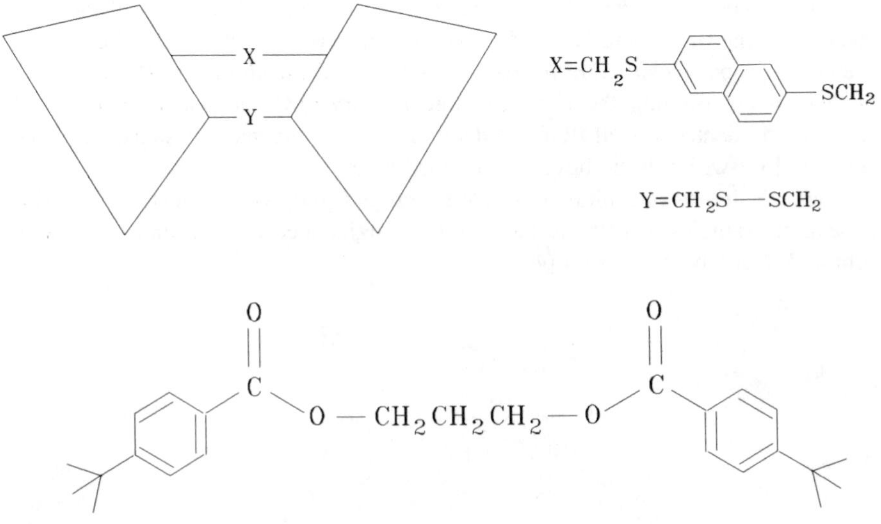

**Figure 10.8**  A schematic view of a very strong complex synthesized by Breslow and co-workers.

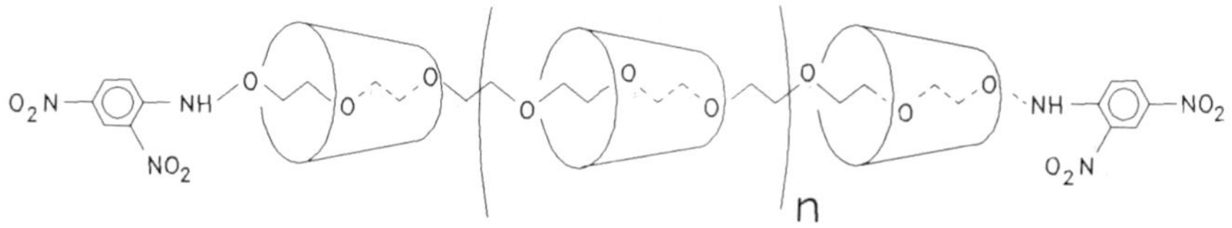

**Figure 10.9** A cyclodextrin complex of a polyrotaxane structure.

of including a metal ion [41] to form a structure similar to the Russian nested doll toys, is an example of CyDs forming a second complexation sphere. A computer model of a part of this impressive supermolecule is presented in color chart 5. An exceptionally good matching of the host and guest in the very strong complex shown in Fig. 10.8 [42] results in stability constants of the order of $10^8$ $M^{-1}$, while the corresponding values for typical complexes with CyDs lie between 10 and 100 $M^{-1}$. A surprisingly simple one-pot reaction yields a rotaxane structure consisting of a polyether thread with numerous CyDs beads, as schematically shown in Fig. 10.9 [43]. It provides another example of the effectiveness of an appropriately chosen self-assembling process. The CyDs complexes, as well as Cram's hemicarcerand **345** [6] and the Whitesides complexes **17**, Scheme 10.1 [5], provide examples of the design of supramolecular complexes with predefined architecture and function. Numerous examples of this kind are described in the literature [1, 44], and the common procedure for a newly synthesized cagelike molecule, such as **132** [45] described in Chapter 4, is to check the size of its cavity for complexing ability.

## 10.4 Mono- and Polylayers, Micelles, Vesicles, and Other Highly Organized Structures [46–48]

To model the structure and operation of organic tissues and to manufacture new material with predefined properties, higher organized structures involving larger numbers of molecules than simple supramolecular complexes have to be considered. According to Tomalia [49], such self-assembling aggregates, called *mesoscopic systems,* form an intermediate step between microscopic molecules or simple complexes with a diameter of ca. 20 nm and the macroscopic systems like cells, cellular networks, and polymers larger than $10^4$ nm. Amphiphilic molecules (like those shown in Fig. 10.10), that is, the molecules with a polar "head" and hydrophobic "tail," are the basic building block of such intermediate aggregates. Understandably, in a polar solvent these molecules arrange themselves in such a way that their hydrophobic parts come close. In a nonpolar one, the opposite arrangement is observed. Various types of aggregates of the amphiphilic molecules in a polar solvent are schematically shown in Fig. 10.11. They are formed above a certain concentration since polar–apolar contacts are energetically disfavored. Micelles consisting of 20–100 molecules are formed if the *critical micellar concentration* is reached. Their apolar interior is capable of transporting hydrophobic molecules such as oils, dyes, and some medications. To solubilize and transport polar proteins

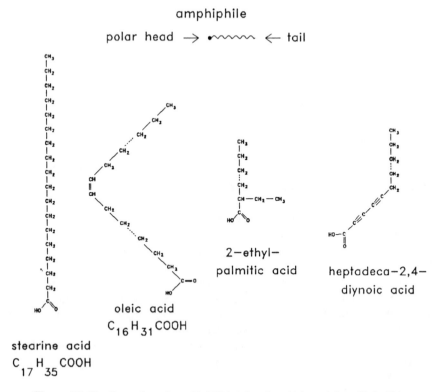

**Figure 10.10**  Examples of amphiphilic molecules. (Adapted from Ref. 49.)

in nonaqueous media, *inverse micelles* with the polar heads pointing inward have been used. Mono-, bi-, or polylayers of amphiphilic molecules can be formed, among others, by dipping an appropriate plate into a polar solvent with a layer of amphiphilic molecules on its surface. The process is depicted in Fig. 10.12. The layered structure obtained in such a way is called a *Langmuir–Blodgett film* [48]. Studies on mono- and polylayers provide considerable information on membranes and liposomal structures [46, 47]. Bilayers of amphiphilic molecules are the main building blocks of cell membranes. The *liposomes* or *vesicles* schematically presented in Fig. 10.11 are enclosed systems formed by one or more bilayers. Their formation is caused by the unfavorable thermodynamics at the end sections of simple bilayer membranes. The liposomes from 15 to 1000 nm are divided into four classes according to their size (1) *small unilamellar vesicles* (SUV), the smallest size limit possible (15 nm for egg lecithin) for phospholipid or other amphiphiles; (2) *intermediate-size unilamellar vesicles* (IUV), with diameters about 100 nm; (c) *multilamellar vesicles* (MLV), with a wide population range (100–1000 nm) formed by five or more layers; and (d) *large unilamellar vesicles* (LUV), with diameters of ca. 1000 nm.

The association of amphiphiles strongly depends on the solvent, temperature,

Schematical presentation of structures
formed by amphiphiles in the solid and solution states
(the hydrophilic regions of each molecule are shaded)

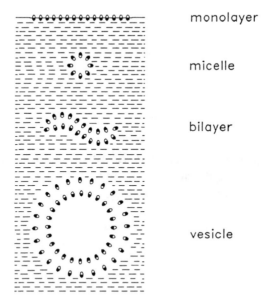

Figure 10.11   Various types of aggregates formed by amphiphilic molecules. (Adapted from Ref. 46.)

## The formation of Langmuir-Blodgett films

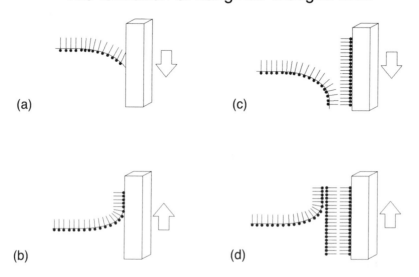

Figure 10.12   Scheme of the formation of Langmuir–Blodgett films. (Adapted from Ref. 49.)

pH, and concentration used. Decher [50] studied various ways of self-organization of a cholesterol derivative. For this derivative he was able to produce two liquid crystalline phases, a monolayer, micelles, or liposomes. According to Ringsdorf [47], there is an analogy between liquid crystals [51] and biological membranes; thus the former can model the behavior of the latter.

As pointed out by Coleman [46], studies of molecular aggregates require the use of experimental techniques different from the physical methods applied in the studies of isolated molecules. Among others they include light scattering, Langmuir balance, measurement of the surface tension and potential, diffraction and spectroscopic methods modified for these studies, atomic force microscopy, etc. The interested reader should consult Refs. 46 and 48.

## 10.5 Some Examples of Highly Efficient Chemical Reactions Based on a Self-Organization Mechanism Which Has Not Been Proposed

The formation of a double-stranded helicate characterized by a positive cooperativity, yielding only starting and final materials with no partially assembled species,

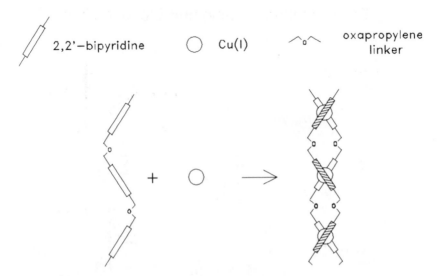

**Figure 10.13** The scheme of formation of double-stranded helicates. (Adapted from Ref. 52.)

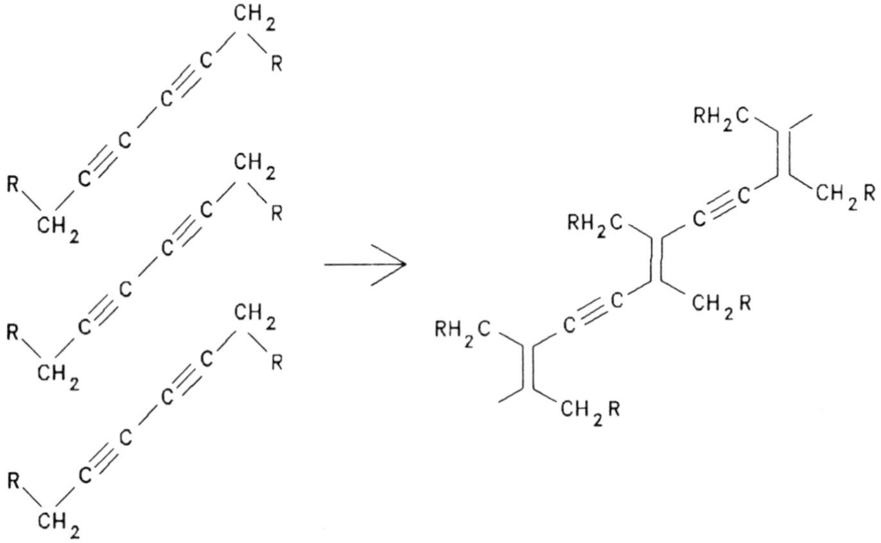

**Scheme 10.2**  An example of cascade reactions based on intramolecular self-assembly involving the formation of irreversible covalent bonds.

is presented in Fig. 10.13 [52]. This reaction is reversible, like most of those involving complex formation. On the other hand, the cascade reactions [53] shown in Schemes 10.2–10.4 are examples of irreversible covalent self-assemblies. Some other examples of this kind involving a very precise matching of the reagents were presented in the Introduction, as well as in Section 9.1, discussing the syntheses of topological molecules. They will also be presented in Chapter 12, dealing with the manufacture of new materials.

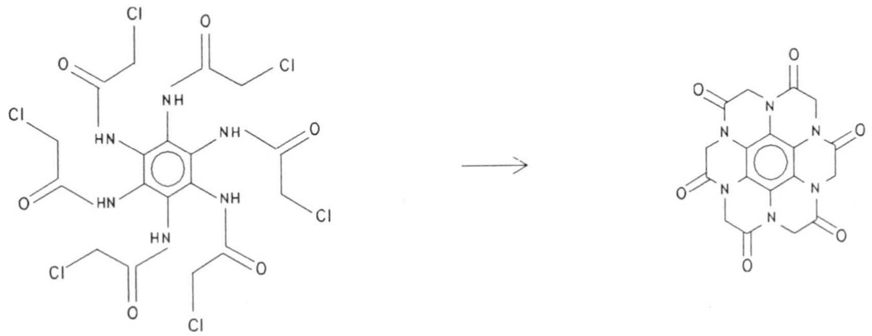

**Scheme 10.3**  An example of cascade cyclization making use of the proper arrangement of molecular fragments.

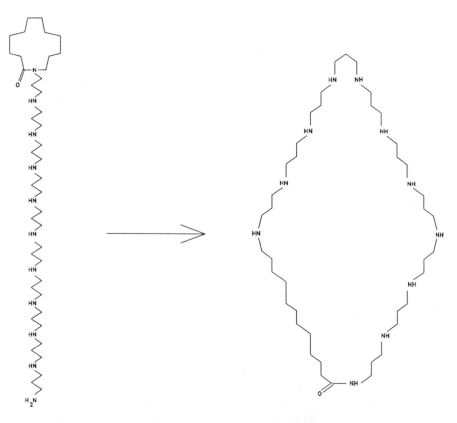

**Scheme 10.4** "Zipper reaction" consisting in 40-atom ring expansion.

# References

1a. J.-M. Lehn, in *Frontiers in Supramolecular Chemistry and Photochemistry*, H.-J. Schneider and H. Dürr, Eds., VCH, Weinheim, 1991.

1b. J.-M. Lehn, *Angew. Chem., Int. Ed. Engl.* **27** (1988) 89.

1c. F. Vögtle, *Supramolekulare Chemie*, Teubner, Stuttgart, 1989.

1d. Ref. 1c, pp. 202–43.

2. *Molecular Association, Including Molecular Complexes*, R. Foster, Ed., Academic Press, New York, 1979; R. Foster, *Organic Charge-Transfer Complexes*, Academic Press, New York, 1969.

3. A. F. Wells, *Structural Inorganic Chemistry*, Oxford University Press, Oxford, England, 1984.

4. T. Tjivikua, P. Ballester, and J. Rebek, Jr., *J. Am. Chem. Soc.* **112** (1990) 1249.

5. J. A. Zerkovski, C. T. Seto, and G. M. Whitesides, *J. Am. Chem. Soc.* **114** (1992) 5473; C. T.

Seto and G. M. Whitesides, *ibid.* **115** (1993) 905; C. T. Seto and G. M. Whitesides, *ibid.* **115** (1993) 1330.

6. D. J. Cram, M. E. Tanner, and R. Thomas, *Angew. Chem., Int. Ed. Engl.* **30** (1991) 1024.

7. *Cyclodextrins and Their Industrial Uses,* D. Duchene, Ed., Editions de Sante, Paris, 1987.

8. J. Szejtli, *Cyclodextrin Techlonogy,* Kluwer Academic Publishers, Dordrecht, 1988.

9. E. Fischer, *Ber.* **27** (1894) 2985.

10. D. E. Koshland, Jr., *Angew. Chem., Int. Ed. Engl.* **33** (1994) 2475.

11. R. Breslow, *Science* **218** (1982) 532, and references cited therein.

12. M. Eigen, *Naturwiss.* **58** (1971) 465.

13. P. R. Sundararajan and V. S. R. Rao, *Carbohydr. Res.* **13** (1970) 351.

14. J. F. Stoddart, First International School of Supramolecular Chemistry, Strasbourg, France, September 16–28, 1990.

15. V. Schurig and H. P. Novotny, *Angew. Chem., Int. Ed. Engl.* **29** (1990) 939.

16. A. Stadler-Szöke and J. Szejtli, *Acta Pharm. Hung.* **49** (1979) 30.

17. S. Li and W. C. Purdy, *Chem. Rev.* **92** (1992) 1457.

18. K. Harata, Chapter 9 in *Inclusion Compounds,* J. L. Atwood and J. E. D. Davis, Eds., Oxford University Press, Vol. 5, 1991, p. 311.

19a. M. Sakurai, M. Kitagawa, H. Hoshi, Y. Inoue, and R. Chujo, *Carbohydr. Res.* **198** (1990) 181.

19b. I. Tabushi and T. Mizutami, *Tetrahedr.* **43** (1987) 1439; C. A. Venanzi, P. M. Canzius, Z. Zhang, and J. D. Bunce, *J. Comput. Chem.* **10** (1989) 1038; F. M. Menger and M. J. Sherrod, *J. Am. Chem. Soc.* **110** (1988) 8606.

19c. J. E. H. Koehler, W. Saenger, and W. F. van Gunsteren, *Eur. Biophys. J.* **15** (1987) 197, 211; *ibid.* **16** (1988) 153.

20. Even if most pharmaceutical and other applications are based on solid complexes, at the site of their action, they are usually dissolved.

21. Y. Inoue, Y. Takahashi, and R. Chujo, *Carbohydr. Res.* **144** (1985) C9–C11.

22. E. L. Eliel, N. L. Allinger, S. J. Angyal, and G. A. Morrison, *Conformational Analysis,* Interscience, New York, 1965, Chapter 3.3.

23. H. Dodziuk and K. Nowinski, *J. Mol. Struct. (THEOCHEM)* **304** (1994) 61.

24. N. L. Allinger, *Adv. Phys. Org. Chem.* **13** (1976) 1; E. Osawa and H. Musso, *Angew. Chem., Int. Ed. Engl.* **22** (1983) 1.

25. N. L. Allinger and Y. H. Yuh, Operating Instructions to the MM2 and MMP2 Programs, updated January 1980, University of Georgia, Athens, GA.

26. Y. Inoue, *Ann. Rep. NMR Spectry.* **27** (1993) 59. This review discusses almost excusively the author's works on the NMR spectra of complexes of the parent CyDs and their derivatives with phenols. It does not take into account a few works documenting separate signals for free and complexed host and guest molecules [27b] and almost totally neglects the NMR manifestations of chiral recognition by CyDs [28, 35,36].

27a. H. Ogino, *New J. Chem.* **17** (1993) 683.

27b. H. Yonemura, M. Kasahara, H. Saito, H. Nakamura, and T. Matsuo, *J. Phys. Chem.* **96** (1992) 5765; M. Watanabe, H. Nakamura, and T. Matsuo, *Bull. Chem. Soc. Jpn.* **65** (1992) 164.

28. H. Dodziuk, J. Sitkowski, L. Stefaniak, D. Sybilska, and J. Jurczak, *J. Chem. Soc. Chem. Commun.* (1992) 207.

29. C. Betzel, W. Saenger, B. E. Hingerty, and G. M. Brown, *J. Am. Chem. Soc.* **106** (1984) 7545; V. Zabel, W. Saenger, and S. A. Mason, *J. Am. Chem. Soc.* **108** (1986) 3664.

30. M. G. Usha and R. J. Wittebort, *J. Am. Chem. Soc.* **114** (1992) 1541.

31. Y. Inoue, T. Okuda, and R. Chujo, *Carbohydr. Res.* **141** (1985) 179; Y. Yamamoto, M. Onda, Y. Takahashi, Y. Inoue, and R. Chujo, *ibid.* **182** (1988) 41.

32. F.-H. Kuan, Y. Inoue, and R. Chujo, *J. Incl. Phenom. Mol. Recogn. Chem.* **4** (1986) 281.

33. D. Sybilska, M. Asztemborska, A. Bielejewska, J. Kowalczyk, H. Dodziuk, K. Duszczyk, H. Lamparczyk, and P. Zarzycki, *Chromatographia* **35** (1993) 637.

34. D. Sybilska and J. Jurczak, *Carbohyd. Res.* **192** (1989) 243.

35. D. Greatbanks and R. Pickford, *Magn. Res. Chem.* **25** (1987) 208.

36. A. F. Casy and A. D. Mercer, *Magn. Res. Chem.* **26** (1988) 765.

37. J. S. Manka and D. S. Lawrence, *Tetrahedr. Lett.* **30** (1989) 7341.

38. P. Job, *Ann. Chim.* **9** (1928) 113.

39. F. Djedaini and B. Perly, "Nuclear Magnetic Resonance of Cyclodextrins, Derivatives, and Inclusion Compounds," in *New Trends in Cyclodextrins and Derivatives,* D. Duchene, Ed., Edition de Sante, Paris, France, 1991.

40. G. Patonay and I. M. Warner, *J. Incl. Phen. Mol. Recogn. Chem.* **11** (1991) 313.

41. F. Vögtle and W. M. Müller, *Angew. Chem.* **91** (1979) 676; S. Kamitori, K. Hirotsu, and T. Higuchi, *J. Am. Chem. Soc.* **109** (1987) 2409.

42. R. Breslow and S. Chung, *J. Am. Chem. Soc.* **112** (1990) 9659.

43. J. F. Stoddart, *Angew. Chem., Int. Ed. Engl.* **31** (1992) 846; A. Harada J. Kamachi, *Nature* **356** (1992) 325; G. Wenz, F. Wolf, M. Wagner, and S. Kubik, *New J. Chem.* **17** (1993) 729.

44. J. Rebek, Jr., *Angew. Chem., Int. Ed. Engl.* **29** (1990) 245.

45. F. Vögtle, J. Gross, C. Seel, and M. Nieger, *Angew. Chem., Int. Ed. Engl.* **31** (1992) 1069.

46. A. W. Coleman and P. Zhang, *Membranes, Liposomes and Supramolecular Chemistry,* Second International School of Supramolecular Chemistry, Strasbourg, France, 1992.

47. H. Ringsdorf, B. Schlarb, and J. Venzmer, *Angew. Chem., Int. Ed. Engl.* **27** (1988) 113.

48. D. A. Tomalia, A. M. Naylor, and W. A. Goddart III, *Angew. Chem., Int. Ed. Engl.* **29** (1990) 138.

49. *Langmuir–Blodgett Films,* G. Roberts, Ed., Plenum Press, New York, 1990.

50. G. Decher, Dissertation, Universität Mainz, 1986, cited by H. Ringsdorf, B. Schlarb, and J. Venzmer, *Angew. Chem., Int. Ed. Engl.* **27** (1988) 113.

51. Liquid crystals are in an intermediate state between liquids, with molecules fully disordered, and solid crystals, in which molecules form equidistant arrays [47].

52. J.-M. Lehn, A. Rigault, J. Siegel, J. Harrowfield, B. Chevrier, and D. Moras, *Proc. Natl. Acad. Sci. USA* **84** (1987) 2565.

53. J. Thomaides, P. Maslak, and R. Breslow, *J. Am. Chem. Soc.* **110** (1988) 3970; V. Enkelmann, *Adv. Polym. Sci.* **63** (1984) 91; U. Kramer, A. Guggisberg, M. Hesse, and H. Schmid, *Angew. Chem., Int. Ed. Engl.* **17** (1978) 200.

# Molecular Modeling

As discussed previously, molecular models only represent certain aspects of the structure of molecules [1, 2], but they are indispensable in the research and development of new materials. At present, mechanical models (Dreiding, CPK, and others discussed in Chapter 1) are being replaced by models generated in computers. According to the Hopfinger definition [3], "Molecular modeling is the generation, manipulation, and/or representation of three-dimensional structures of molecules and associated physicochemical properties." There are four methods of computer model generation:

1. Building from fragments of standard geometry taken from a database or by two-dimensional sketching on the screen. The former procedure is carried out by clicking on icons representing molecular fragments on the screen and showing, also by clicking, the place where they have to be added. Such models without a subsequent refinement are rather imprecise. For instance, their bond angles may be in error by as much as 10°. Subsequent minimization of these structures by means of QC, MM, or MD methods is a common procedure. Molecular builders and the strategy on which they are based are discussed in Ref. 4. When the model is built by sketching, the molecular skeleton is drawn in the screen plane. The third coordinates of the atoms are obtained from some user hints. The latter are exemplified by showing the way in which the dodecahedrane model is built in the PCMODEL program [5a].

2. Using X-ray databases to represent the whole molecule or its fragment is especially practical for proteins and other large molecules. As mentioned in Chapter 2, the Cambridge Structural Data Base, containing more than $10^5$ structures

of small and middle-sized molecules as of 1994 [6], and the Brookhaven Protein Data Base [7] are the sources of such information. The use of crystal structures can also be questionable, since they differ from the solution structures (see the discussion in Chapter 2), where the effects to be studied primarily operate.

3. Theoretical calculations—QM, MM, MD. Their merits and limitations were discussed in Chapter 3.

4. Using information on distance geometry obtained by NOE, as briefly presented in Chapter 2. This method, which is widely applied in association with theoretical calculations for proteins and other biopolymers [8], lies outside the scope of this monograph.

It should be stressed that, due to rotational isomerism, the number of conformations that can be assumed by larger molecules is so large that they become intractable even for today's computers [4]. For instance, taking into account three possible orientations around a $C_{sp^3}$—$C_{sp^3}$ bond yields almost 60,000 rotamers for $C_{11}H_{24}$ having a chain of ten such bonds. The rotamer number increases to several million for the chain hydrocarbon $C_{16}H_{34}$ consisting of only 50 atoms. The problem of finding the global minimum is theoretically unsolvable. However, several approaches, including systematic or random conformational search or simulated annealing, are applied to obtain acceptable solutions [9].

As mentioned before, numerous program packages for molecular modeling are on the market. Some of them are listed in Ref. 10, but no critical evaluation is given. In addition to the molecule building part, they always contain an MM program using various force field parameters. As discussed in Section 3.3, numerous FF have been proposed in the literature. Some of them are considered to be general, but specific FF have been proposed for certain classes of compounds such as proteins [11]. Carrying out some kind of, usually semiempirical, QC is also possible with more sophisticated programs. Larger program packages for workstations, such as that by Tripos with the graphic front-end program SYBYL [5b], the one by BIOSYM with the front-end program INSIGHT [5c], QUANTA CHARMm by Molecular Simulations [5d], or HyperChem [5e], allow one to perform MD calculations or Monte Carlo dynamic simulations. In these cases an automatic inclusion of solvent molecules to study solution behavior is possible. Molecular modeling programs on PCs such as PCMODEL [5a] are very user-friendly. They can and are used by synthetic chemists to model problems since, as will be shown, structure building and minimization requires very little knowledge. In addition to numerous advantages, it also creates some serious problems, since users often have no understanding of limitations of the programs they use. Arriving at a local minimum instead of the global one, the force-field dependence of the results of the calculations provide examples of problems that can be unrecognized by an inexperienced program user. Some larger program packages [5b–5e] allow one to choose the FF or minimization procedure used. This implies a much deeper understanding of the program operation, but does not always solve the problems.

As discussed, several program packages for workstations exist. They have mostly been developed to model proteins. However, for the modeling of smaller organic molecules several simpler programs for personal computers can be used, including PCMODEL, Chem3D/Plus, HYPERCHEM, ALCHEMY [5], etc. Here the model building will be briefly exemplified by showing an application of the PCMODEL program [5a]. To our knowledge, the same strategy is applied in other modeling packages. The program operates in three modes. In the first one, consisting of modeling, called INPUT, the molecular model can be created by (1) reading in the input file constructed earlier. Another possibility is (2) to draw the planar carbon skeleton of the molecule under study with a subsequent positioning of certain atoms below or above the screen defining the plane of the drawing by clicking the "in" or "out" icon. Alternatively, a frequently used molecular fragment, such as a six-membered ring in a chair or boat conformation, can be invoked by clicking the appropriate icon. Then hydrogen atoms can be added by clicking the "H add" icon, and an atom replacement can be carried out by clicking the appropriate icon and the position of the substitution, etc. Double bonds are created by drawing the line between two carbon atoms positions twice. Then they have to be marked as bearing $\pi$ atoms.

It should be stressed that one can learn modeling only by practicing it with a program. For a person familiar with computers the learning process is fast and fun. However, it cannot be learned by studying books only. Therefore, modeling will be illustrated here by building the dodecahedrane **8** model presented in Fig. 11.1.

The planar model defining the connectivity scheme was drawn first in the order shown by numbering the atoms; that is, the positions of the atoms 1, 2, 3, 4, 5, 1 were clicked first to draw a five-membered ring, and the next rings were added to it. For instance, by clicking the positions of atoms 5, 6, 7, 8, 4, the second ring was built. The clicking of the atomic positions 8, 9, 10, 3 was sufficient for the formation of the next ring, etc. Then atoms denoted by "x" were moved below the screen

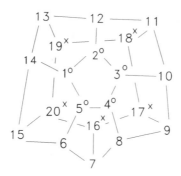

**Figure 11.1**   The building of the dodecahedrane **8** model using the PCMODEL program. The order of numbering exemplifies the order in which the atom positions are clicked. The planar model is converted into a three-dimensional one by moving atoms denoted by "x" below the screen, while those denoted by "o" were moved above the screen. It must be stressed that in this case even such an inaccurate procedure suffices to obtain a beautiful $I_h$ symmetrical structure after energy minimization. However, it is not necessarily the case, and the structure obtained in the minimization process can correspond to a local, and not the global, minimum.

by clicking the "in" icon, while those denoted by "o" were moved above it by clicking the "out" one. However, using this program, it is not possible to build models with reasonable spatial structure for complicated molecules like α-cyclodextrin **14a.** When the dodecahedrane model is built, the clicking of the "minimization" and "energy" icons in the subsequent MINIMIZATION mode yielded a regular structure of $I_h$ symmetry (see Chapter 4 for symmetry notation). The minimization can involve simple MM calculations. Alternatively, it can consist of a combination of MM and semiempirical QC, as is the case in the PCMODEL program.

In more sophisticated program packages MM calculations are usually only the first step, followed by QC and/or MD simulations. After the minimization, the DISPLAY mode of PCMODEL provides rather modest possibilities for structure visualization. For instance, the PCMODEL version at our disposal allows only for stick models to be displayed. Other programs have much more sophisticated display capabilities. In addition to rotation of the model, much more sophisticated than that available in PCMODEL, those programs usually allow for stick, space-filling CPK molecular representations or a mixed model representation with atoms displayed as small spheres. They are all shown in color charts 1, 3, 4, 5, and 6. There is also a possibility of observing molecular fragments by zooming and clipping. Various modes of presentation of α-CyD with stick, ball-and-stick, and ball models obtained using the SYBYL program package are presented in color chart 1. The ball models of α-, β-, and γ-CyDs **14a–c,** respectively, with the ball radii proportional to the van der Waals ones, clearly reveal the increasing size of the cavities responsible for the inclusion complex formation presented in color chart 6. Modeling packages more sophisticated than PCMODEL enable one to present and analyze not only the spatial molecular structure but also electrostatic potentials, hydrophobicity of molecular surfaces, etc.

As discussed in Chapter 3, molecular dynamics and Monte Carlo simulations provide information on the time evolution of the system under investigation. On the basis of these calculations, modeling packages enable one to visualize the changes in molecular geometry during the simulation. In addition to graphs describing the changes of an analyzed property (such as energy, a torsional angle, or a nonbonded distance), molecular models corresponding to geometries at selected time intervals are projected. Moreover, a movielike animation can be displayed, reproducing molecular geometry at chosen time steps.

Molecular modeling forms the basis of *computer-assisted molecular design* (CAMD). Two main approaches are applied in rational drug design involving CAMD. In the first, based on quantitative structure–activity relations (QSAR), discussed in the next chapter, no knowledge of the operation of the drug is assumed. In the second, a single mechanism of the biological activity under investigation, namely, the formation of the receptor–ligand complex, is a premise of the studies. In relatively few cases is the receptor X-ray structure known and its active site recognized. Then various drug molecules are docked into the site interactively by the researcher or automatically by the program. The calculated energy of stabilization of the drug–receptor complex enables one to understand the drug operation contributing to the development of a novel, effective drug. Numerous studies of an

enzyme dehydrofolate reductase (DHFR), discussed in Ref. 9a, in particular, modeling [12] and the X-ray structure studies of their enzyme–inhibitor complexes [13],
are examples of such an approach. Another much more demanding strategy aims at
modeling the active site of a receptor on the basis of the structure and activity of
known drugs. Such an approach is illustrated by a study of norfloxacin-derived
drugs [14]. A more potent drug is next designed on the basis of this knowledge. As
discussed in the next chapter, such studies are seldom published in view of the
competitiveness of the drug market. Other examples of the application of molecular
modeling are shown in Chapters 2, 3, 10, and 12.

Let us look more closely at the work of Ohta and Koga [14] on an improvement
of norfloxacin **353** ($X = $ CH). This drug has opened a new age in quinolone
antibacterials and oral anti-infectives, since in addition to potent activity against
both Gram-positive and Gram-negative bacteria, it exhibits a low incidence of
resistant bacteria, an oral effectiveness against systemic infections, a metabolic
stability, and a low oral toxicity [15]. On the basis of the study of 71 compounds of
the general formula **354,** a quantitative structure–activity relation (QSAR) function
(see the next chapter) was proposed relating the minimum inhibitory concentration
(MIC) of the drug (in mol/l) against *Escherichia coli* NIHJ JC-2 bacteria with the

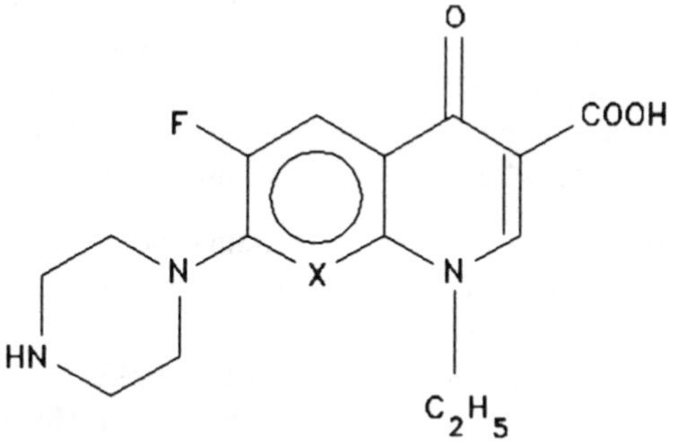

X=CH norfloxacin

**353**

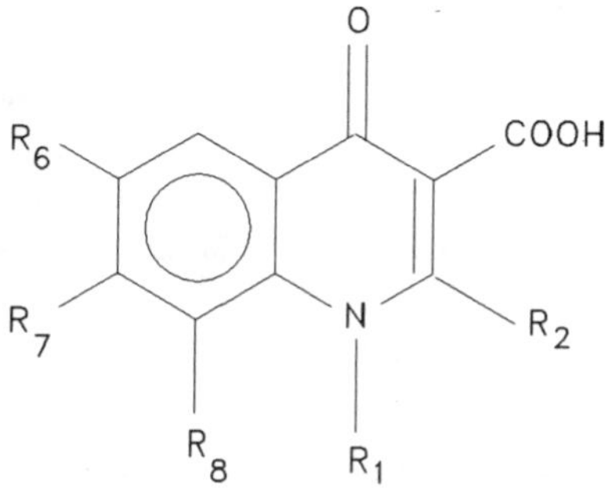

**354**

length of the $R_1$ substituent, the maximum width of $R_8$, hydrophobic parameters of $R_6$, $R_7$, and $R_8$ (related to the transport of compounds into the active site), Taft–Kutter–Hansch steric parameters of $R_6$, Swain–Lupton–Hansch inductive electronic parameters of $R_6$, $R_7$, $R_8$, etc. It was also clear that steric factors of the $R_1$, $R_6$, and $R_8$ substituents are important for the biological activity. However, a new highly active compound, ciprofloxacin **354** ($R_1$ = cyclopropane, $R_7$ = —N⬭NCH$_3$, $R_8$ = H), was synthesized next, for which the prediction of activity based on the latter QSAR relation failed. Then a series of modeling calculations was carried out for a larger series **353,354** consisting of (1) building of molecular models from fragments, (2) optimizing the structures by means of MM minimization, and (3) QC minimization of the MM-minimized structures using GAUSSIAN82, AM1, and CNDO/2 programs. The comparison of the calculated structures with the experimentally determined pharmacologic activity revealed that:

1. Not only the length but also the orientation of the $R_1$ substituent is of importance.
2. There are two optimum regions for increasing the drug activity: one corresponding to the region occupied by an $N_1$ cyclopropyl group above the plane of the quinolone ring. The other region for $R_1$ aromatic ring substituents with $R_6$ = F and various $R_1$, $R_7$, and $R_8$ substituents corresponds to the region occupied by a fluorine atom and a hydroxyl group at the *para*-position of the $N_1$-phenyl.
3. Moreover, there are two regions unfavorable to proper receptor binding so that their occupation leads to a reduction in the activity. One of them corresponds to the region below the plane of the quinolone ring and the other, to the *meta* position of the $N_1$-phenyl group above the quinolone ring plane.

Such a study allowed one to obtain a certain insight into the structure of the active site of the receptor interacting with this class of drugs and was of importance in modifications of the norfloxacin.

As discussed in the next chapter, the successful applications of CAMD in pharmaceutical and other industries as a rule are not published. However, the abundance of modeling packages in the market and the prices for which they are sold are indicative of their usefulness. Other CAMD applications are less known. Among others, they deal with studies of the properties of solid surfaces to be used in developing new heterogeneous catalysts as well as novel organic materials for electronics and optoelectronics.

## References

1. R. Hoffmann and P. Laszlo, *Angew. Chem., Int. Ed. Engl.* **30** (1991) 1.

2. T. Gund and P. Gund, in *Molecular Structure and Energetics*, Vol. 4, J. Liebman and A. Greenberg, Eds., p. 319, VCH Publishers, New York, 1987.

3. A. J. Hopfinger, *J. Med. Chem.* **28** (1985) 1133.

4. J. Sadowski and J. Gasteiger, *Chem. Rev.* **93** (1993) 2567.

5a. PCMODEL is distributed by Serena Software, Dr. Kevin E. Gilbert, P.O. 3076, Bloomington, IN 47402.

5b.  SYBYL and ALCHEMY are distributed by TRIPOS Associates, 1699 S. Hanley Road, Suite 303, St. Louis, MO 63144.

5c.  Insight/Discover is distributed by BIOSYM Technologies, Inc., 9685 Scranton Road, San Diego, CA 92121-2777.

5d.  QUANTA CHARMm is distributed by Molecular Simulations Inc., 16 New England Executive Park, Burlington, MA 01803-5297.

5e.  HyperChem is distributed by Autodesk, Inc., 2320 Marinship Way, P.O. Box 399, Sausalito, CA 94965-9950.

5f.  Chem3D/Plus is distributed by Cambridge Scientific Computing, Inc., Dr. Stewart Rubinstein, 875 Massachusetts Avenue, Suite 41, Cambridge, MA 02139.

 6.  F. H. Allen, S. Bellard, M. D. Brice, B. A. Cartwright, A. Doubleday, H. Higgs, T. Hummelink, B. G. Hummelink-Peters, O. Kennard, W. D. S. Motherwell, J. R. Rodgers, and D. G. Watson, *Acta Crystallogr. Sect. B* **B35** (1979) 2331.

 7.  Protein Data Bank, Chemistry Department, Building 555, Brookhaven National Laboratory, Upton, NY 11973.

 8.  W. Braun, *Q. Rev. Biophys.* **19** (1987) 115; K. Wüthrich, *Acc. Chem. Res.* **22** (1989) 36; A. K. Ghose and G. M. Crippen, "Distance Geometry Approach to Modeling Receptor Sites," in *Comprehensive Medicinal Chemistry,* C. Hansch, P. G. Sammes, and J. B. Taylor, Eds., Pergamon Press, New York, 1990.

9a.  H. Frübeis, R. Klein, and H. Wallmeier, *Angew. Chem., Int. Ed. Engl.,* **26** (1987) 403.

9b.  P. Amara, D. Hsu, and J. E. Straub, *J. Phys. Chem.* **97** (1993) 6715.

10.  D. B. Boyd, "Compendium of Software for Molecular Modeling," in *Reviews in Computational Chemistry,* VCH, New York, 1990.

11.  J. Weiner, P. A. Kollman, D. A. Case, U. C. Singh, C. Ghio, G. Alagona, S. Profeta, Jr., and P. Weiner, *J. Am. Chem. Soc.* **106** (1984) 765; H. A. Scheraga, *Adv. Phys. Org. Chem.* **6** (1968) 103.

12.  C. D. Selassie, Z. X. Fang, R. L. Li, C. Hansch, T. Klein, R. Langridge, and T. B. Kaufman, *J. Med. Chem.* **29** (1986) 621; C. Hansch, B. A. Hathaway, Z. Guo, C. D. Selassi, S. W. Dietrich, J. M. Blaney, R. Langridge, K. W. Volz, and T. B. Kaufman, *ibid.* **27** (1984) 129.

13.  J. T. Bolin, D. J. Filman, D. A. Matthews, R. C. Hamlin, and J. Kraut, *J. Biol. Chem.* **257** (1982) 13650; D. J. Wolz, D. A. Matthews, R. A. Alden, S. T. Freer, C. Hansch, T. B. Kaufman, and J. Kraut, *ibid.* **257** (1982) 2582; D. J. Filman, J. T. Bolin, D. A. Matthews, and J. Kraut, *ibid.* **257** (1982) 13663.

14.  M. Ohta and H. Koga, *J. Med. Chem.* **34** (1991) 131, and references cited therein.

15.  B. Holms, R. N. Brogden, and D. M. Richards, *Drugs* **30** (1985) 482.

# 12

# Present and Future Applications

As will be shown in this chapter, a few of the new and unusual molecules and supramolecular systems described in this book have found practical applications. However, a deeper understanding of the interrelationship between molecular shape and electron distribution gained by studying such systems allows one to plan the synthesis of molecules with predetermined desired properties on the basis of molecular modeling involving CAMD. Few of them have yet found practical applications, and much work is still needed to put into operation recent concepts of novel medicines and nonstandard administration methods as well as new ideas in molecular electronics and optoelectronics. With the last two, miniaturization is a must for bringing about faster data processing accompanied by a decrease in the weight and power used, leading to higher efficiency and a considerable price decrease.

A comparison of the effectiveness of natural and synthetic systems seems instructive [1a]. The capacity of the human brain equals ca. $10^{15}$ bits. In 1989 one needed $2 \times 10^6$ magnetic disks of 60 MB to mimic brain operation. (1 byte corresponds to 8 bits.) Every such disk used 50 watts; thus, 0.1 GW was necessary for the whole system, while only a few watts are used by the brain. Semiconductor memory devices cost at that time $10 for a 1 Mbit chip. Thus a memory corresponding to that of a human brain would cost $10^9$ [1b]. Admittedly, the latter is less reliable, but it is highly cost effective. Similarly, to reproduce the sensitivity of a human eye, one had to use $10^8$ photomultipliers costing ca. $100 each. Thus $10 billion was needed to build a device with a similar performance to that of the human eye.

At present the ultimate goal in miniaturization (which lies far in the future) are single-molecule, single-supermolecule electronic devices, or their supermolecular assemblies. Basic ideas on the structure requirements and operation of such devices will be discussed here briefly. Some examples of molecules that have been proposed for use in electronics in the future will also be shown. Industrial applications of optoelectronics are even more distant, but the payoff they seem to offer is enormous. Therefore, massive efforts are directed to the study of new materials with pronounced nonlinear properties. In the following the applications in the pharmaceutical industry and agrochemistry will be presented first. Then some organic molecules and complexes pertaining to chemionics (i.e., electronics based on nonstandard molecules) will be discussed. A few examples of molecules exhibiting considerable nonlinear behavior will conclude this chapter.

## 12.1 Pharmaceutical and Agrochemical Applications

### 12.1.1 Rational Drug Design

The development of a new drug is a tedious task that takes 8 to 15 years. Similarly, 8 to 10 years may be needed to develop a new animal health product, herbicide, fungicide, or insecticide. The development of a new medicine involves costly syntheses and testing of biological activity of ca. 15,000 molecules; thus any method allowing one to decrease this number is of importance. In recent years CAMD (introduced in the preceding chapter), based on an effective drug–receptor complex, has gained in importance in drug design. It is based on the rapid development of computer capabilities, especially those associated with the graphical representation of molecular shapes and surfaces. The first monograph on the application of computers in drug design was published as early as 1979 [2]. Since then, quite a few commercial programs developed for this purpose have been proposed. Sybyl [3] and Quanta CHARMm [4] (see Chapter 11 for a short description of molecular modeling) are probably the most frequently used in the pharmaceutical industry. It should be stressed that only a few successful industrial applications of CAMD have been documented in the literature, since the market is very competitive, and the results, which are proprietary of pharmaceutical companies, are published only if interest in a specific class of molecules as a potent drug has waned.

The mechanism of a drug's action is not known in most cases, and empirical quantitative structure–activity relationships (QSAR) are a standard tool in drug design. In the Hansch approach [5] of QSAR, biological activity, expressed as the reciprocal of the concentration $C$, that brings about a biological response is correlated with certain molecular descriptors

$$\log(1/C) = a_o + \sum_i b_i D_i$$

The responses may describe anesthetic, depressant, hypnotic, or anticonvulsant activity, as well as acute lethal toxicity of the drug under investigation. As follows from the Gupta reviews [6, 7], separate correlations for numerous drug classes have been proposed. One of the main molecular descriptors determining drug action involves the drug lipophilicity (i.e., hydrophobicity enabling its penetration through cell membranes). For instance, the crossing of the blood–brain barrier is the crucial step predetermining the desired activity of all narcotic analgesics. In a series of compounds, a contribution of the substituent $X$ to the drug lipophilicity can be accounted for by comparing the partition coefficient $P_X$, usually for an $n$-octanol/water system, with the corresponding value for the parent compound $P_H$ [8]. The dependence on log $P$ is generally parabolic since the molecules with too high a lipophilicity could be trapped in the cell membranes. The role of hydrophobicity in the interaction of various drugs with the receptor site is much more diversified. Electron donating or accepting properties are usually taken into account by using various $\sigma$ Hammett constants [9] or quantum-mechanically derived charges or energies. The ability to form hydrogen bonds and/or charge-transfer complexes was shown to influence the activity of certain classes of drugs [6, 7]. Topological indices [10] and steric parameters involving the Taft $E_s$ constants and the Verloop length and width of a substituent [11] are also used for certain groups of drugs as well as the molar refractivity and polarizability of a substituent [9]. It should be stressed that the QSAR correlations between the above parameters and biological activity critically depend on the accuracy of biological data, which, in turn, depend on several factors. The list of parameters most frequently used in QSAR is given and discussed in Ref. 6.

Most of the QSAR studies are based on an assumption of rigid molecular structures in well-defined conformations, not only for a drug, but also for its receptor. As a rule, this is not the case. The structures determined by means of X-ray analyses are not necessarily the same as those in solutions where the molecules are more flexible than they are in the crystalline state. Even in the latter state, in most cases there is a considerable degree of flexibility, which can influence the drug action. Thus its nonrigidity is an important factor that must be taken into account.

The application of molecular modeling to drug design was exemplified in the preceding chapter by the work of Ohta and Koga [12] on the improvement of norfloxacin **353** ($X = $ CH).

## 12.1.2 New Modes of Drug Administration

Today not only new, highly potent drugs are being developed, but also the way in which they are administered is changing. As described in Chapter 10, an inclusion of a drug sensitive to light and water into the CyD cavity prevents its decomposition. It also increases the solubility of drugs in water. In addition, the slow release of the drug from the cavity makes possible a significant change in the way it is applied. For instance, instead of administering the drug every 4 hours, it can be given once in 2 or 3 days. This method is not only more convenient for a patient and the medical personnel, but it also maintains a more uniform level of the medicine in the organism during the treatment.

The demand to use only enantiomerically pure chiral drugs, which has not been fulfilled yet, will eventually remove the 50% impurity from the racemic drugs marketed today. As discussed in Chapters 5 and 10, the most effective enantiomeric separation is achieved by chromatographic techniques using CyDs as stationary or mobile phases. The same methods can be used for the separation of isomer mixtures [13].

In recent years, the strategy used by to develop plant-protecting agents and other agrochemicals has also changed drastically. For instance, instead of using DDT or other relatively simple chemicals that kill insects and pollute, much gentler techniques have been adopted that consist in using juvenile hormones that block the insect's transformation from a larva to a pupa. Another possibility being explored is the use of the insect's sexual attractors to prevent mating and reproduction. Hormones like **355** and attractors do not create environmental problems, since they easily decompose to nonpolluting organic molecules.

355

## 12.2 Molecular Electronics—Chemionics

As mentioned, the miniaturization of electronic devices speeds up information processing, as well as decreasing weight, price, and power used. Organic molecules are the prospective building blocks of future electronic devices, since they are relatively inexpensive and easy to process materials. In addition, due to the specific role played by the carbon atom in organic chemistry, it is relatively easy to tune the properties of organic materials by modifying known molecular structures or developing new ones. These are important advantages of organic molecules, but the most important point is that some organic compounds exhibit enhanced electric responses over a wide frequency range and ultrafast response times. The high laser-damage thresholds they exhibit is of importance for optoelectronics applications. Finally, the realm of organic molecules that may find industrial application is huge and ever-growing. Actually, the situation is similar to that in drug design. Prospective molecules are too numerous even to check their properties; thus CAMD calculations are indispensable in developing new materials.

Today mesoscopic systems 1 nm to 1 $\mu$m in size [14] are being prepared for use in electronic devices. In the future, devices consisting of a single molecule or supermolecule will be desirable. The latter approach seems especially promising, since supramolecular systems offer the possibility for a precise adjustment of their properties by design. According to Lehn [15], "The processing of molecular information via molecular recognition events implies a passage from the molecular to the supramolecular level. By endowing photo-, electro-, or ionoactive components with recognition features, one may be able to design programmed molecular systems that undergo self-assembly into organized and funcional (photonic, electronic, ionic) supramolecular devices, and these are in turn likely to reach nanometer dimensions. Thus the chemistry of supramolecular devices, *chemionics*, is the chemistry of molecular recognition-directed, self-organized, and functional entities of supramolecular nature." Some sophisticated host receptors recently synthesized and their aggregates, especially Langmuir–Blodgett films, that were briefly presented in Chapter 10, illustrate the versatile possibilities of synthetic chemistry in obtaining systems with predetermined desired properties. In the following, a few systems promising specific applications will be shown. Then the strategy for obtaining logical elements for use in computers as memory, logic, and amplification elements will be presented.

## 12.2.1 Molecular Wires

Molecular wires, proposed by Lehn [15, 16], are the essential components of electronic circuitry connecting different parts of an electronic device and enabling electron flow to take place between them. Such a role could be played by caroviologens **356** possessing a conjugated polyene chain for electron conduction, terminal electroactive and hydrosoluble groups for reversible electron exchange, and a length sufficient for spanning typical molecular supporting elements such as a monolayer or bilayer membrane. As schematically shown in Fig. 12.1, they could also be incorporated into a vesicle functioning as a continuous, transmembrane electron channel. Various modifications of the end groups and chain can be made [16] to modulate their functioning. For instance, coupling with photoactive groups could give photoresponsive electron channels; differentiation of the end group could result in charge separation and signal transfer devices; etc.

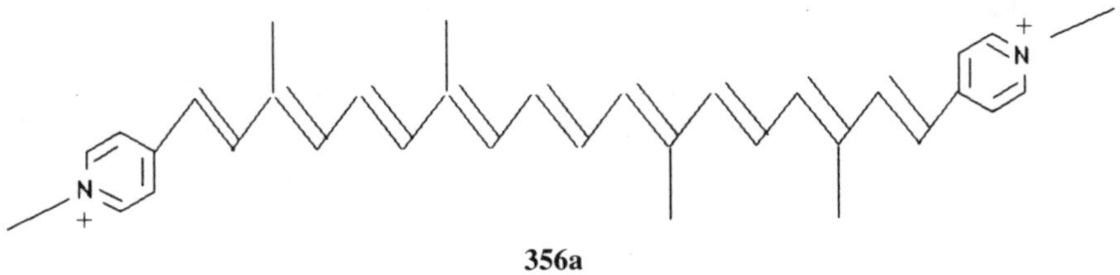

**356a**

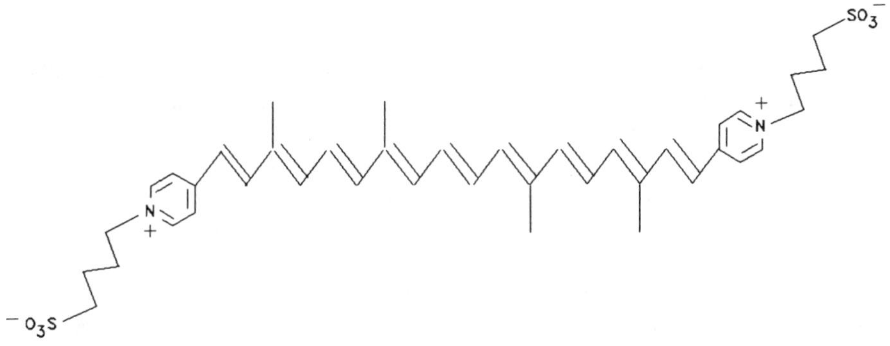

**356b**

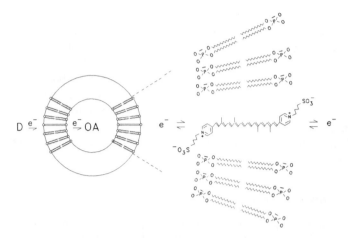

**Figure 12.1** Schematic representation of transmembrane electron transfer by a zwitterionic caroviologen (in this case **356b**). *OA* is an oxidating agent; *D*, a reducing one; and the caroviologen is incorporated in the bilayer membrane of a phospholipid vesicle. (Adapted from Ref. 15b.)

## 12.2.2 Ionic Transmitters or Detectors

Signal and information storage, processing, and transfer could be carried out by molecular or supramolecular ionic devices built from receptor, reagent, or carrier molecules capable of handling inorganic or organic ions. Similarly to the corresponding electronic devices, they function via highly selective recognition, reaction, and transport processes. Both molecular (charge, size, shape, and structure) and supramolecular (binding energy, strength, and selectivity) features confer on ionic devices a very high information content. However, they are expected to react more slowly than electronic devices. Supramolecular assemblies of macrocyclic compounds **357a, 357b** in the form of the tubular liquid crystalline mesophase **357c** could form the basis of the device acting as an ionic channel [15, 16].

## 12.2.3 Organic Metals

Surprisingly, some organic salts and their charge-transfer complexes exhibit values as high as $10^3$ (ohm cm)$^{-1}$ of their room-temperature conductivity [17]. This value is almost equal to that of the metal bismuth. Some organic molecules forming charge-transfer complexes that have been proposed as a basis for prospective organic conductors are **95, 358, 359** [18, 19]. Metallomacrocycles such as phthalocyanines **360** have also been studied with this purpose in mind. The most promising organic conductor today seems to be the salt of the charge-transfer complex of tetrathiafulvalene (TTF) **358a** ($R = H$, $X = S$) acting as a donor (D), with tetracyanoquinodimethane (TCNQ) **359a** ($R = H$) playing the acceptor (A) role. It is interesting that the desired conducting properties are exhibited only if the molecules crystallize in layers forming the segregated TTF-TCNQ stacks DD∥AA. The mixed stacks D·A·D·A do not show any significant conductivity [20].

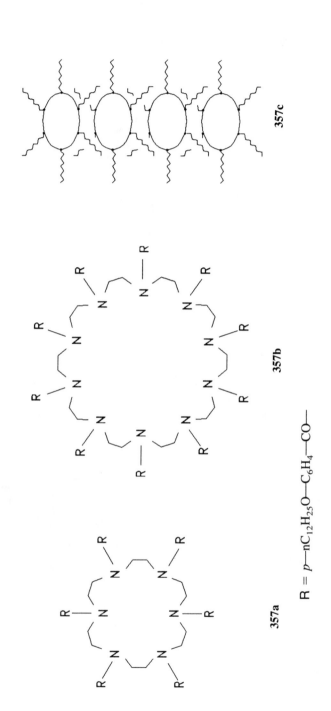

357a

357b

357c

R = p—nC$_{12}$H$_{25}$O—C$_6$H$_4$—CO—

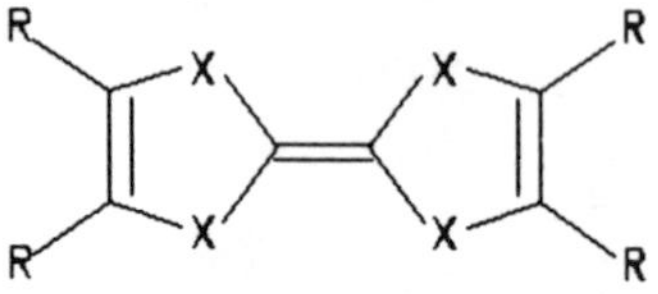

R=H,Me; X=S,Se

358a

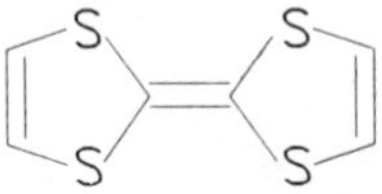

358a1

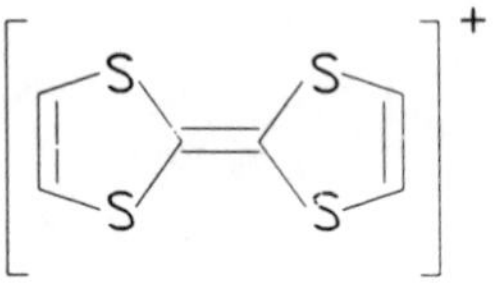

358a2

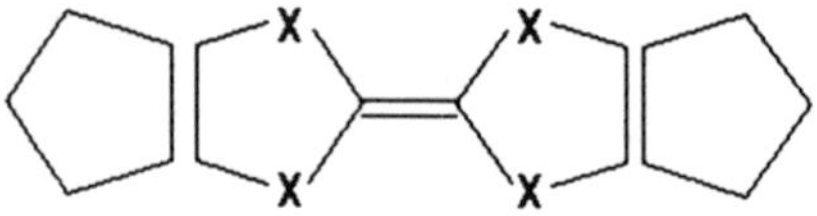

X=Se,Te

358b

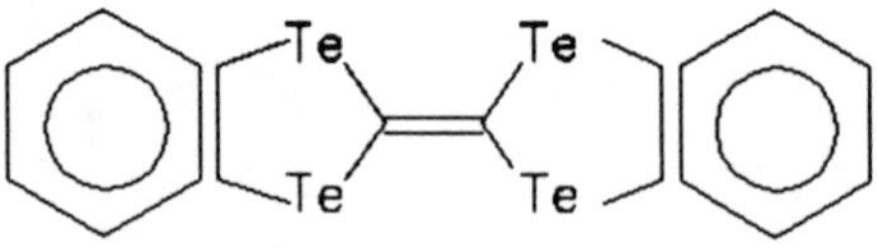

358c

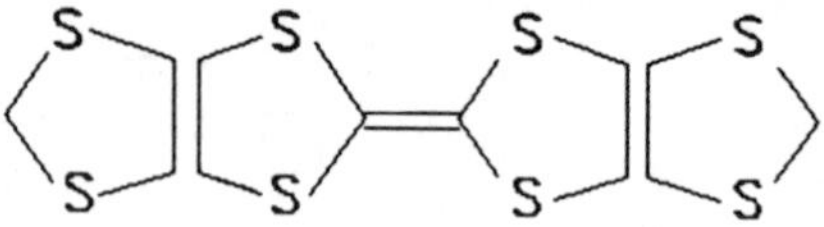

358d

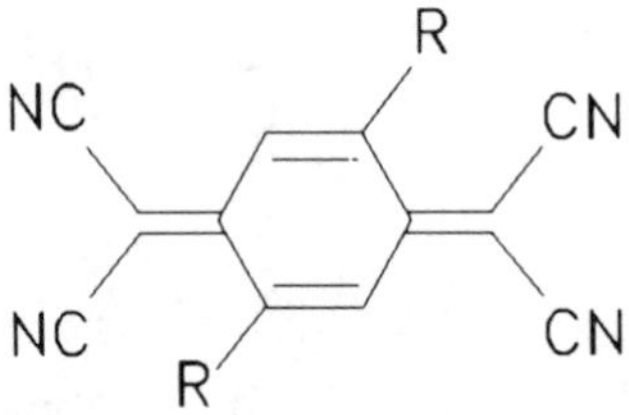

R=H,Me

359a

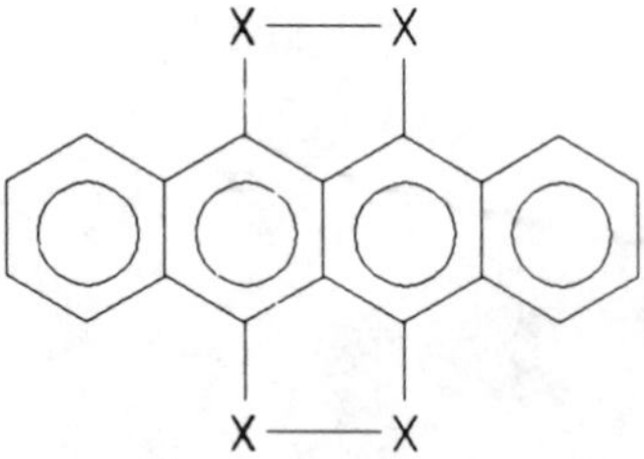

X=S,Se

359b

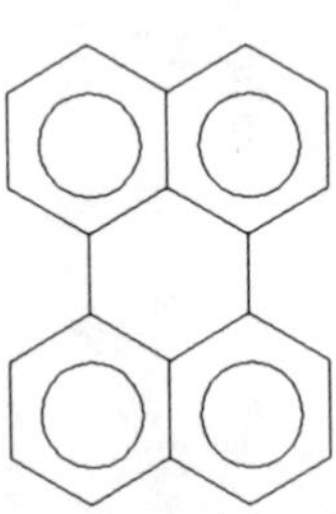

359c

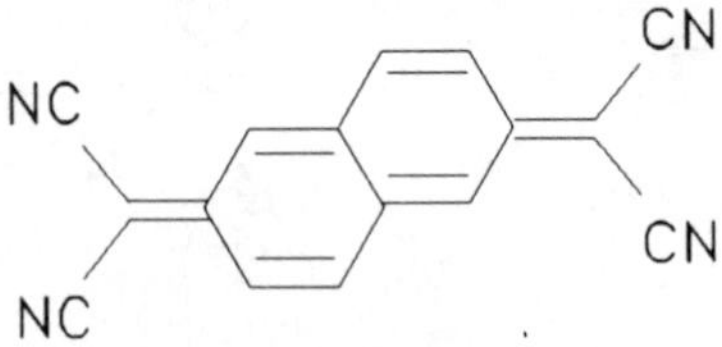

359d

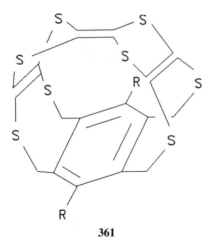

M=metal

**360**

The molecules **358** have planar tetrathiafulvalene units [21], while the cage compound **361** [19] possesses the unit highly distorted from planarity.

**361**

As mentioned in Chapter 8, $C_{60}$ offers the possibility of obtaining superconducting materials at very low temperatures [22]. If one could obtain superconducting materials at room temperature, e.g., on the basis of $C_{60}$, then huge savings could be expected by eliminating losses due to long-distance energy transfer.

### 12.2.4 Organic Switches—The Molecules for Memory, Logic and Amplification

#### 12.2.4.1 The Concepts [23, 24]

As mentioned, organic molecules for use as conductors often exist in two forms: the nonconducting proconductor such as TTF **358a1** (PC) and the conductor (C) tetra-thiafulvalenium anion **358a2.** Reduction or oxidation brings about the exchange of one of these forms into the other. Aviram examined the possibility of constructing a molecule that can serve as a proconductor or a conductor when placed between two metallic electrodes. The HOMO orbitals of the proconductor molecule is located below the Fermi level [25] of the metal. It contains two electrons; thus no tunneling can take place. The current between the electrodes can flow only if a bias voltage is applied. On the contrary, the partial filling of HOMO in the conductor form allows electrons to be transfered between the electrodes when a small bias is applied. Therefore, the difference between PC and C forms can be formulated in the following way: The first one acts as an insulator up to a threshold voltage, while there is practically no threshold voltage for the latter form.

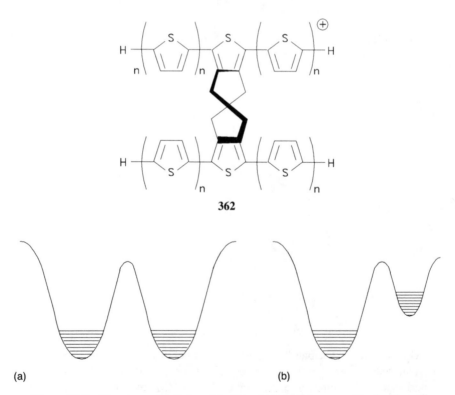

**Figure 12.2**  The energy levels in a (a) symmetric and (b) asymmetric double well.

To create a molecule that could serve as an electronic device, Aviram proposed a system **362** consisting of the conducting and proconducting units bridged by a spirojunction. This molecule provides a double-well potential for an electron [presented in Fig. 12.2(a)] that can "jump" between the wells due to the tunneling effect introduced in Chapter 3. The frequency of this transition is characteristic of the specific molecule involved. Fine tuning this frequency is the purpose of structural engineering, which should in future enable one to construct molecules having desired properties. The tunneling frequency can be controlled, for instance, by adjusting the separation distance between the PC and C parts of the device across the $\sigma$ bridge, by modifying the symmetry of the double-well potential or by regulating the overlap (through-space or through-bridge) between the molecular orbitals of the conductor and proconductor. Obviously, the overlap is the biggest when the two systems are parallel and the smallest when they are orthogonal. If the latter orientation does not suppress the tunneling, then the molecule **362** can form a building block for future computers.

After the creation of the double-well asymmetry [Fig. 12.2(b)], it can serve as a bit in a binary system. There an electron localized in one well would represent the state "one," while its localization in the second well would correspond to the "zero" state. Checking the conductivity from two separate terminals would function as information reading. The double-well asymmetry can result from the application of an electric field along the $z$ axis, which is determined by the spiro center and the centers of the C and PC units.

When the system **362** was proposed by Aviram [23], its desired length was set at ca. 500 pm. Since then two oligomers of this kind have been synthesized [26]. However, the technological problems seem to sharpen the requirements, and in Ref. 24 Aviram puts the desired length of these molecules at 2500 pm.

The functional details of the molecular switch [23, 24] based on **362** will not be discussed here. Similarly, the method to develop logical elements ("OR", "AND" and "EXCLUSIVE OR" gates) on the basis of such molecular switches described in the same articles lies outside the scope of this book. It suffices to say that the concepts for building monomolecular devices for memory, logic, and amplification have been formulated. Their realization requires overcoming enormous technological problems, expanding our knowledge about molecules at interfaces. The development of new surface-science analytical methods, that of surface binding techniques such as self-assembly [27], and recent spectacular advances in nanolithographic techniques [28] are the first steps in the long route, which will revolutionize our lives by building much more efficient means of communication.

## 12.2.4.2 Examples of Organic Molecules as Prospective Molecular Switches

In the system **362** proposed by Aviram, an electron transfer between two states formed the basis of operation of an electronic device. Photochemical processes trigger the transformation from one electronic state to the other in the molecules **363b,c** (M, M1 are metal atoms) [29] and **364** [30]. Stolarczyk and Piela [31] proposed a molecule with a few stable states as a prospective switch. They discussed the conditions under which an electric field would trigger transitions between these electronic states. A chiroptical switch is presented in Scheme 12.1 [32], while intramolecular metal atoms hopping between two calixarene moieties could be used in another type of molecular switch [33].

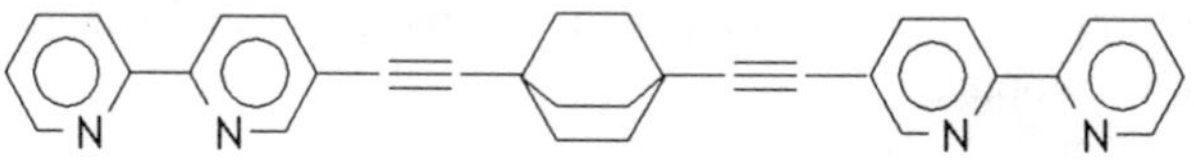

**363a**

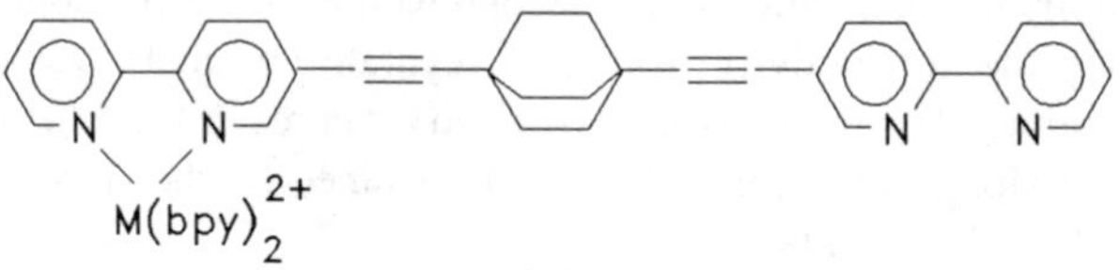

**363b**

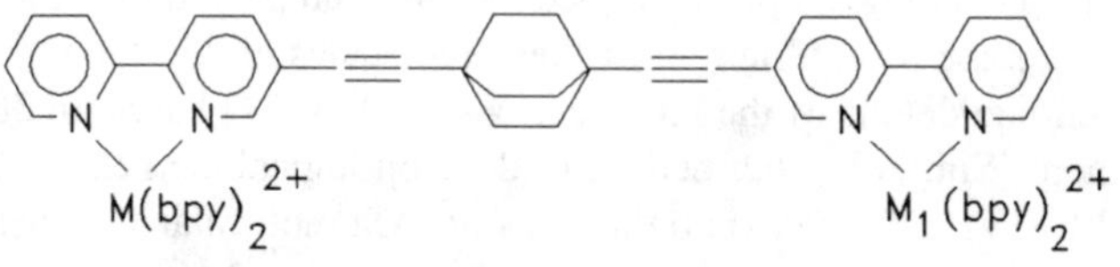

**363c**

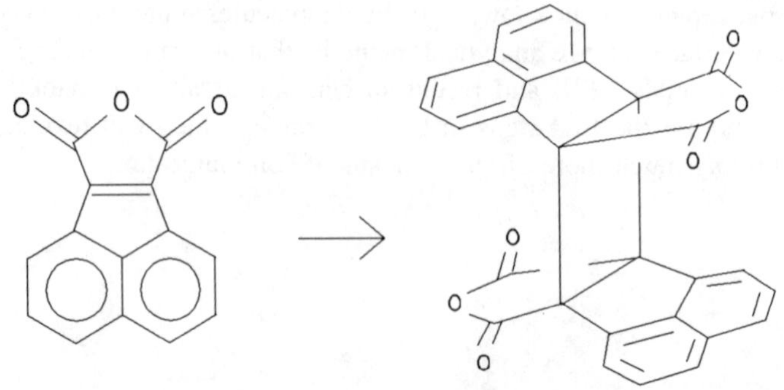

**364a**                          **364b**

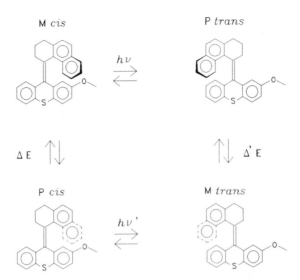

**Scheme 12.1** A chiroptical switch proposed in Ref. 30.

## 12.2.5 Organic Systems for Future Use in Optoelectronics

The passage from electronics to optoelectronics is a still greater challenge. Optical communication seems to offer revolutionary high data-transmission rates that are free from electrical interference. Dynamical image processing and optical computing will in future transform everyday life. Prospective applications in optoelectronics are based on nonlinear optical phenomena [34, 35]. The electric field **E** of an electromagnetic wave propagating through a medium induces a polarization **P** and a magnetization **M** as a result of the motion of the electrons and nuclei of this medium in response to the applied field. Under certain assumptions, the induced polarization can be expanded into a power series in the electric field of the light wave of the form

$$P_i(\mathbf{r},\,t) = \epsilon_0 \left( \sum_j \chi_{ij}^{(1)}\, E_j(\mathbf{r},\,t + \sum_{j,k} \chi_{i,j,k}^{(2)}\, E_j(\mathbf{r},\,t)\, E_k(\mathbf{r},\,t) \right.$$

$$\left. + \sum_{j,k,l} \chi_{ijkl}^{(3)}\, E_j(\mathbf{r},\,t)\, E_k(\mathbf{r},\,t)\, E_l(\mathbf{r},\,t) + \ldots \right)$$

where the indices $i$, $j$, $k$, and $l$ denote the $x,y,z$ components of the vector and tensor quantities and the $\chi_{ij}^{(n)} \ldots$ coefficients are the $n$th-order electric susceptibilities. The first coefficient is involved in the linear-optical effects, while other terms describe nonlinear-optical effects of the $n$th order.

Strong nonlinear effects are exhibited by both the amphiphilic and non-amphiphilic molecules shown in Scheme 12.2. *Ortho-* and *para*-substituted aromatic structures in which the substituents stabilize mesomeric structures with a partial separation of charge (Scheme 12.3) also exhibit strong nonlinear effects [36].

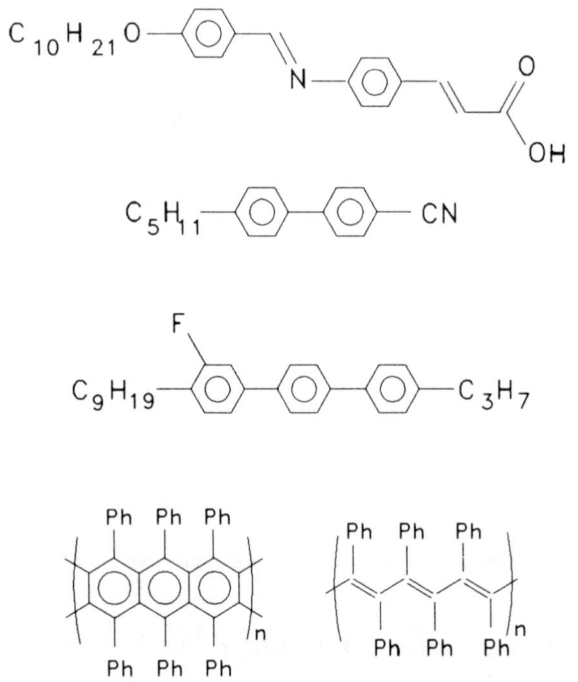

**Scheme 12.2**   Some examples of molecules exhibiting strong nonlinear effects.

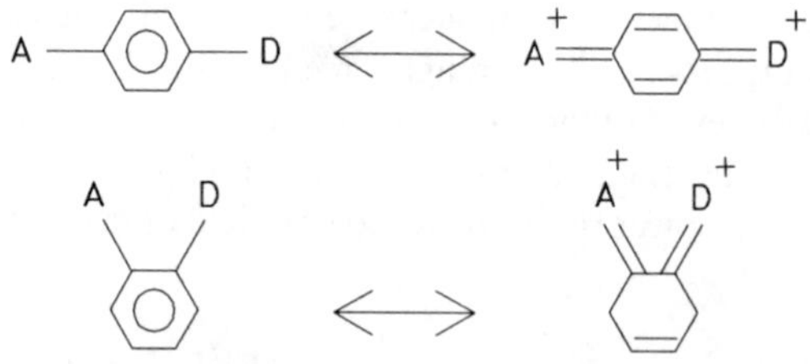

**Scheme 12.3**   Mesomeric structures of *ortho*- and *para*-disubstituted benzenes exhibiting strong nonlinear effects.


## 12.3 Concluding Remarks

Only a few prospective applications of unusual organic molecules could be shown in this chapter, since the field is so wide. Moreover, at present it is not clear which molecules claimed to find practical applications will succeed. This area presents enormous technological problems that will require interdisciplinary approach to solve. Many technological bottlenecks have to be overcome until molecular devices consisting of a single molecule or a single supermolecule can become a reality. The rapid development of synthetic methods in organic and organometallic chemistry and in their structure determination on one hand and the rise of computers and CAMD methods on the other are prerequisites for future expansion of novel applications of organic and organometallic molecules. Studies of such systems will not only bring payoffs from industrial applications, but they will also enrich our knowledge of intra- and intermolecular interactions in which molecular shapes and electron distributions play a decisive role. Such studies will also contribute to our understanding of the operation of living organisms.

This book is devoted to novel trends in the conformational analysis of organic molecules. The field is blooming, and in the author's opinion it involves a majority of the most important areas of organic chemistry. New exciting molecules and supramolecular systems are constantly being synthesized, their structure determined, and theoretical calculations used to rationalize or predict their properties. Theoretical methods also play ever-increasing role in obtaining custom-made molecules or supermolecules for practical applications. The understanding of self-association processes governed by the matching of molecular shapes and electrostatic charge distributions will undoubtedly contribute to our understanding of the operation of living organisms. The field is undergoing exciting developments: Every month brings interesting findings, and no limits can be foreseen.

## References

1a. D. Haarer, *Angew. Chem., Int. Ed. Engl., Adv. Mater.* **28** (1989) 1544. However, it is not clear whether the analogy between a human brain and the computer assumed by Haarer is fully justified.

1b. In 1994 1 GB, (that is, 1 gigabyte) magnetic disc cost ca. $1000. Thus, when this book was being written memory corresponding to that of a human brain would cost $10^8$, an order of magnitude less than 5 years ago. This example shows the rapid development of computers and the long way still to go if we wish to match the efficiency of living organisms.

2. E. C. Olson and R. E. Christoffersen, *Computer-Assisted Drug Design*, ACS Symposium Series 112, American Chemical Society, Washington, DC, 1979.

3. SYBYL, distributed by TRIPOS Associates, 1699 S. Hanley Road, Suite 303, St. Louis, MO 63144.

4. QUANTA CHARMm distributed by Molecular Simulations, Inc., 16 New England Executive Park, Burlington, MA 01803-5297.

5. C. Hansch and T. Fujita, *J. Am. Chem. Soc.* **86** (1964) 1616; C. Hansch, *Acc. Chem. Res.* **2** (1969) 232; C. Hansch, in *Drug Design*, E. J. Ariens, Ed., Academic Press, New York, 1971, Vol. 1, p. 271.

6. S. P. Gupta, *Chem. Rev.* **87** (1987) 1183.

7. S. P. Gupta, *Chem. Rev.* **89** (1989) 1765.

8. T. Fujita, J. Iwasa, and C. Hansch, *J. Am. Chem. Soc.* **86** (1964) 5175.

9. C. Hansch and A. Leo, *Substituent Constants for Correlation Analysis in Chemistry and Biology*, J. Wiley, New York, 1979.

10. S. P. Gupta, P. Singh, and M. C. Bindal, *Chem. Rev.* **83** (1983) 633; L. B. Kier and L. H. Hall, *Molecular Connectivity in Chemistry and Drug Research*, Academic Press, New York, 1976.

11. A. Verloop, W. Hoogenstraaten, and J. Tipker, *Med. Chem.* **11** (1976) 165; A. Verloop, W. Hoogenstraaten, and J. Tipker, in *Drug Design*, E. J. Ariens, Ed., Academic Press, New York, Vol. 7, p. 165; A. Verloop and J. Tipker, *Pharmacochem. Libr.* **10** (1987) 97.

12. M. Ohta and H. Koga, *J. Med. Chem.* **34** (1991) 131, and references cited therein.

13. S. Li and W. C. Purdy, *Chem. Rev.* **92** (1992) 1457.

14. D. A. Tomalia, A. M. Naylor, and W. A. Goddard III, *Angew. Chem., Int. Ed. Engl.* **29** (1990) 138.

15. J.-M. Lehn, *Angew. Chem., Int. Ed. Engl.* **27** (1988) 89.

16. J.-M. Lehn, in *Frontiers in Supramolecular Chemistry and Photochemistry,* H.-J. Schneider and H. Dürr, Eds., VCH Publishers, Weinheim, 1991, p. 1.

17. D. O. Cowan and F. M. Wright, *Chem. Eng. News* (July 21, 1986) 28.

18. T. J. Marks, *Angew. Chem., Int. Ed. Engl.* **29** (1990) 857.

19. S. S. Shaik and M.-H. Whangbo, *Inorg. Chem.* **25** (1986) 1201.

20. S. S. Shaik, *J. Am. Chem. Soc.* **104** (1982) 5328.

21. J. Rörich, P. Wolf, V. Enkelmann, and K. Müllen, *Angew. Chem., Int. Ed. Engl.* **27** (1988) 1377.

22. J. F. Stoddart, *Angew. Chem., Int. Ed. Engl.* **30** (1991) 70.

23. A. Aviram, *J. Am. Chem. Soc.* **110** (1988) 5687.

24. A. Aviram, *Int. J. Quantum Chem.* **42** (1992) 1615.

25. The Fermi level is the energy level separating the occupied and unoccupied states.

26. J. M. Tour, R. Wu, and J. S. Schumm, *J. Am. Chem. Soc.* **113** (1991) 7064.

27. P. E. Labinis, J. J. Hickman, M. S. Wrighton, and G. M. Whitesides, *Science* **245** (1989) 845.

28. H. I. Smith, K. Ismail, W. Chu, A. Yen, Y. C. Ku, D. A. Antoniadis, and M. L. Schattenburg, in *Proceedings of the Conference: Molecular Electronics, Science and Technology,* A. Aviram, Ed., United Engineering Trustees, New York, 1990, p. 107, and other papers.

29. F. Vögtle, M. Frank, M. Nieger, P. Belser, A. von Zelewsky, V. Balzani, F. Barigelletti, L. De Colla, and L. Flamigni, *Angew. Chem., Int. Ed. Engl.* **32** (1993) 1643.

30. R. D. Rieke, G. O. Page, P. M. Hudnall, R. W. Arhart, and T. W. Bouldin, *J. Chem. Soc., Chem. Commun.* (1990) 38.

31. L. Z. Stolarczyk and L. Piela, *Chem. Phys.* **85** (1984) 451.

32. B. L. Feringa, W. F. Jager, and B. de Lange, *J. Am. Chem. Soc.* **113** (1991) 5470.

33. F. Ohseto and S. Shinkai, poster B-43 at the XVIII International Symposium on Macrocyclic Chemistry, Enschede, the Netherlands, June 27–July 2, 1993.

34. G. H. Wagniere, *Linear and Nonlinear Properties of Molecules,* VCH, Weinheim, 1993.

35. J.-M. Andre and J. Delhalle, *Chem. Rev.* **91** (1991) 843.

36. D. J. Williams, *Angew. Chem.,* **96** (1984) 637.

# Index

**259**